Sword & Shield of Zion

To the men and women, past and present,
of the Israel Air Force—the true
Rocks of Israel

SWORD & SHIELD OF ZION

The Israel Air Force in the Arab–Israeli Conflict, 1948–2012

DAVID RODMAN

sussex
ACADEMIC
PRESS
Brighton • Portland • Toronto

2 4 6 8 10 9 7 5 3 1

First published 2013 in Great Britain by
SUSSEX ACADEMIC PRESS
PO Box 139, Eastbourne BN24 9BP

and in the United States of America by
SUSSEX ACADEMIC PRESS
920 NE 58th Ave Suite 300
Portland, Oregon 97213-3786

and in Canada by
SUSSEX ACADEMIC PRESS (CANADA)
8000 Bathurst Street, Unit 1, PO Box 30010, Vaughan, Ontario L4J 0C6

British Library Cataloguing in Publication Data
A CIP catalogue record for this book is available from the British Library.

Library of Congress Cataloging-in-Publication Data
Rodman, David, author.
Sword and shield of Zion : the Israel Air Force in the Arab–Israeli conflict, 1948–2012 / David Rodman.
pages cm
Includes bibliographical references and index.
ISBN 978-1-84519-583-0 (h/b : alk. paper)
1. Israel. Hel ha-avir—History. 2. Israel—History, Military.
3. Israel—Military policy. 4. Arab–Israeli conflict—Aerial operations, Israeli. I. Title.
UG635.I75R63 2013
358.40095694—dc23

2012046589

MIX
Paper from responsible sources
FSC® C013056
FSC www.fsc.org

Typeset & designed by Sussex Academic Press, Brighton & Eastbourne.
Printed by TJ International, Padstow, Cornwall.
Printed on acid-free paper.

Contents

Acknowledgments

I would be remiss if I did not mention the people who contributed to making this monograph a reality. First, I would like to recognize the many men and women of the Israel Air Force (IAF), past and present, who shared their thoughts and experiences with me during my frequent visits to the air force. Among retired and reserve personnel, I would especially like to express my appreciation to Brigadier General Asaf Agmon, Brigadier General Itzhak Amitay, Lieutenant Colonel Amos Cohen, Lieutenant Colonel Danny Grossman, Brigadier General Ephraim Segoli, and Dr. Tal Inbar for providing much helpful information. The many active duty personnel with whom I have had the chance to speak cannot be named for reasons of national and personal security. I have had to content myself with recognizing some of them in the notes by the initials of their first names. Second, I would never have had the opportunity to meet all of these men and women were it not for the warm hospitality offered to me by the Israel Air Force Center Foundation (IAFCF) in the persons of Executive Director Simcha Salach, Executive Assistant Deana Weinrich, and Sharona Haviv. The foundation is the charitable arm of the Israel Air Force Center (IAFC). Known as the "force behind the force" for the many ways in which it supports the air force, the center also serves as a point of contact between the IAF and outsiders like myself.

I would like to thank Efraim Karsh and Rory Miller, in their capacity as editors, for publishing my article, "The Israel Air Force in the 1967 and 1973 Wars: Revisiting the Historical Record," *Israel Affairs*, Vol. 16, No. 2 (April 2010). A portion of this article, revised and adapted, appears in the Introduction (Part II) of this volume, while the remainder constitutes Chapter 1. The journals division of the Taylor & Francis Group, *www.tandfonline.com*, has kindly granted me permission to reproduce my article. Barry Rubin and Yeru Aharoni, in their capacity as editors, deserve thanks, too, for publishing my articles, "Israel's National Security Doctrine: An Introductory Overview," *MERIA Journal*, Vol. 5, No. 3 (September 2001) and "Unmanned Aerial Vehicles in the Service of the Israel Air Force: 'They Will Soar on Wings Like Eagles,'" *MERIA Journal*, Vol. 14, No. 3 (September 2010). Portions of the former article, revised and adapted, constitute the Introduction (Part I) of this monograph. The latter article, revised and adapted, constitutes Chapter 4 of this volume. *MERIA Journal* has kindly given me permission to reproduce my articles. And I would also like to thank Leo Haber, in his capacity as editor, for publishing my article, "Beyond the Battlefield: The Humanitarian Operations of the Israel Defense Forces," *Midstream*, Vol. 52, No. 1 (January–February 2006). This article, revised and adapted, constitutes part of Chapter 3 of this book. *Midstream* has kindly given me permission to reproduce my article as well.

The State of Israel's Government Press Office (National Photo Collection) and my good friend, Israeli aviation enthusiast David Weinrich, who has undoubtedly forgotten more about aircraft than I will ever know, kindly granted their permission for me to use the photographs that appear in this book. David was also gracious enough to answer all of my many questions about aircraft, even the stupid ones, with a smile on his face! Finally, I would like to thank Editorial Director Anthony Grahame and the staff at Sussex Academic Press for showing faith in this work.

It goes without saying, of course, that any errors of substance and/or style in this volume are entirely my responsibility.

List of Abbreviations

AAA	Anti-Aircraft Artillery
ARM	Anti-Radiation Missile
CAS	Close Air Support
C^2	Command and Control
C^3I	Command, Control, Communications, and Intelligence
C^4ISR	Command, Control, Communications, Computers, Intelligence, Surveillance, and Reconnaissance
IADS	Integrated Air Defense System
IAF	Israel Air Force
IAFC	Israel Air Force Center
IDF	Israel Defense Forces
PLO	Palestine Liberation Organization
POL	Petroleum, Oil, Lubricants
PGM	Precision-Guided Munition
RMA	Revolution in Military Affairs
SAM	Surface-to-Air Missile
SAR	Search and Rescue
SAR	Synthetic Aperture Radar
UAV	Unmanned Aerial Vehicle

List of Illustrations

Illustrations appear between chapters 3 and 4. All are reproduced courtesy of the Israeli government and/or the individual photographers.

Preface

The late Israeli Foreign Minister Abba Eban, a highly respected career diplomat and an outstanding orator, once remarked that the State of Israel had not experienced a single day of peace during its history. Known for his crisp summations of the Israeli predicament, Eban's rhetorical skills did not let him down on this occasion. Israel, as even a casual observer of the Middle East scene would certainly be aware, has been engaged in hostilities with its Arab neighbors from the moment of its birth. If it has not been entangled in interstate wars with the surrounding states, then it has been caught up in asymmetrical wars and counterinsurgency campaigns involving nonstate entities. It has also regularly been engaged in special operations warfare.

Airpower has been an integral component of almost all of these bouts of fighting. Consequently, the Israel Air Force (IAF) has accumulated as much battle experience as any air force in the world since the Second World War, and far more than most of them. And it has chalked up more outstanding accomplishments than any other air force. Its record in air-to-air combat is unrivaled in the jet age. The virtual annihilation of the Egyptian, Jordanian, and Syrian air forces on the ground at the beginning of the 1967 Six-Day War and the similarly complete destruction of the Syrian integrated air defense system (IADS) at the outset of the 1982 Lebanon War constitute seminal events in the history of air warfare. The IAF has also been a pioneer in applying airpower to asymmetrical wars and counterinsurgency campaigns, as demonstrated by its efficient eradication of Hizbullah's long-range rocket force in the early days of the 2006 Second Lebanon War and its equally efficient elimination of numerous terrorist operatives during the second intifada (or the so-called al-Aqsa Intifada).

The primary purpose of this monograph is to enumerate the IAF's contribution to Israeli national security through a description and analysis of its part in the state's interstate and asymmetrical wars, counterinsurgency campaigns, and special operations. The book also examines the IAF's contribution to national security with respect to ground-based air defense, space-based reconnaissance, and humanitarian relief operations. To this end, the Introduction, which is divided into two parts, reviews in broad strokes a few of the most salient national security variables that have influenced the development and employment of Israeli airpower, as well as the IAF's combat history. It also defines a few important airpower-related terms. Chapter 1 examines the IAF's performance in two interstate maneuver wars—the Six-Day War and the 1973 Yom Kippur War. Chapter 2 scrutinizes its performance in one interstate and two asymmetrical attrition wars: the 1969–70 War of Attrition, the Second Lebanon War, and the 2008–9 Gaza War (or Operation Cast Lead). Chapter 3 examines the IAF's performance in counterinsurgency campaigns, special operations warfare, and humanitarian operations. Chapter 4 probes one aspect of the air force's own version of the Revolution in Military Affairs (RMA), its development and employment of unmanned aerial vehicles (UAVs). Chapter 5 reviews the IAF's burgeoning capabilities in the realms of ground-based air defense and space-based reconnaissance. Chapter 6 describes the infrastructure—particularly air bases and the squadrons housed at them—that undergirds the IAF's striking power. The Conclusion advances some generalizations about the IAF's past and future contributions to Israeli national security. Finally, a short appendix surveys the evolution of the IAF's fighter-bomber inventory.

A few caveats about this book are in order here. First, it does not purport to offer a comprehensive description and analysis of the IAF's part in all of Israel's interstate wars, asymmetrical wars, counterinsurgency campaigns, special operations, and humanitarian operations. Nor does it furnish a complete account of the air force's experience with UAVs, ground-based air defense, and space-based reconnaissance. Rather, it focuses selectively on those events and developments in which the air force played an especially conspicuous part. Furthermore, it does not offer a blow-by-blow description and analysis of each of these events and developments; instead, the volume focuses more broadly on major themes related to the IAF's part in them. Second, the use of technical terminology, while unavoidable in a discussion about airpower, has been kept as simple as possible for the following reason: the author, who is not an airpower specialist himself, intends this monograph to be accessible to a lay audience. Third, numbers—for example, as in number of sorties flown in a war, as in number of aircraft shot down in a war, or as in number of anti-aircraft emplacements destroyed in a war—often vary from source to source, sometimes quite significantly. In cases of disagreement, this volume has adopted those numbers that are either most plausible or most readily endorsed by experts.

One last issue: this book went to press just before a major flare-up in fighting between Israel and Hamas (as well as its allied terrorist organizations), sparked by the latter's increasingly aggressive rocket attacks; therefore, Operation Pillar of Defense, Israel's name for the latest conflict, is not addressed in this monograph. Nevertheless, two preliminary points can be made at this juncture in late 2012: (1) Israel appears to have achieved one of its three limited aims in the campaign, the crippling of Hamas' (and its allies') rocket-launching capabilities, through an effective combination of air strikes and active/passive anti-rocket defenses; and (2) whether its other two limited aims—a long-term cessation of hostilities across its southern frontier and the reinforcement of deterrence—have been met will be determined by the way events unfold in the coming months and years.

DAVID RODMAN
Dix Hills, New York

Introduction (Part I)
Israeli National Security and Airpower

A state's national security doctrine spells out the means by which the state exploits its military, diplomatic, economic, social, cultural, and other assets—as well as the means by which it compensates for deficiencies in these areas—to protect and promote its national interests. The development and employment of a state's airpower—at least in those states where airpower is not simply an affectation intended to impress friends and foes—is, of course, linked to its national security doctrine. The State of Israel—a state whose airpower is decidedly not "for show"—is certainly no exception to this rule.[1] Most of the variables incorporated into Israel's national security doctrine, however, have had only tenuous connections, if any at all, to the state's airpower. A few of the military-related variables, on the other hand, have clearly had a significant effect on the development and employment of Israeli airpower.

Consequently, it is first advantageous to examine briefly how these variables have exerted an influence on the state's national security doctrine across the past six plus decades, in order to get a sense of the broader military context in which Israeli airpower has evolved over time. These variables include strategic depth and defensible borders, allocation of manpower, quality versus quantity (with respect to both manpower and arms), modes of warfare, and proliferating threats.

Strategic Depth and Defensible Borders

Israel won its 1947–49 War of Independence, actually acquiring considerably more land than originally assigned to it under the terms of the 1947 United Nations Partition Resolution (for British Mandate Palestine), but it nevertheless emerged from the fighting with very problematic frontiers. Extremely long and largely flat, they could not be adequately defended by the fledgling Israel Defense Forces (IDF), as demonstrated by the routine ease with which even untrained Arab terrorists slipped into Israeli territory to inflict mayhem on the civilian population during the early years of statehood. Moreover, Israel had no strategic depth. At its narrow waist, the state's width measured a mere nine miles; and, even in the north and south, its width did not exceed more than a few dozen miles. All of Israel's major population centers, industrial assets, and military bases, then, were potentially within easy reach of Arab armies and terrorist organizations.

This geographical position made a great impression on Israel's defense planners. It quickly led them to conclude that the state could not afford to "host" either an interstate war—an asymmetrical war was not in the cards at this juncture of the state's history—or an insurgency on its territory. The latter occurrence, they reasoned, would inevitably result in extensive damage to Israeli society, not only in terms of the infliction of physical costs, but also in the infliction of psychological costs. And the former, they thought, could well undermine the very existence of the state. Such thinking gave rise to the idea that fighting must be transferred to Arab territory to the greatest extent possible as soon as possible, certainly in the case of interstate war.

This idea, in turn, had profound implications for the IDF's operational and tactical principles, its force structure, and its organizational culture. Suffice it here to say that Israel's geographical conundrum from 1949 to 1967 forms part of the explanation for its emphasis on preventive and preemptive interstate war during these years.[2] Unlike many other states, which have borders that either make it possible for them to prevent invaders from penetrating into their interiors (e.g., Switzerland) or the territorial depth for their own armies to fall back, regroup, and eventually expel the invaders from their interiors (e.g., Russia), Israel inside its pre-1967 borders possessed neither of these luxuries. Therefore, it fought a preventive war in 1956 and a preemptive war in 1967.

With regard to insurgency, Israel's geographical position from 1949 to 1967 prompted its emphasis on retaliation. The IDF had neither the human nor matériel resources to mount an effective perimeter defense, to seal the state's borders against terrorists bent on murder, [...]otage, and theft. Hence, Israeli defense planners decided [...]e needed the "cooperation" of its Arab neighbors to [...]maintain quiet along its frontiers, so it attempted to com[...]stem the tide of infiltration by inflicting costs on them [...]iatory actions. Likewise, Israel sought to put an end to border skirmishing that involved the Arab states themselves through similar retaliatory actions.

The outcome of the 1967 Six-Day War radically altered the geographical status quo in the Arab–Israeli conflict. Not only had Israel completely pulverized the Egyptian, Jordanian, and Syrian armies, but it had also captured large tracts of Egyptian, Jordanian, and Syrian territory. It captured the Sinai and Gaza from Egypt, Judea and Samaria from Jordan, and the Golan from Syria. These territorial acquisitions provided the state with a measure of strategic depth for the first time in its history, particularly vis-à-vis Egypt in the south and Jordan in the east. Israel's major population centers, industrial assets, and military bases no longer remained within easy reach of Arab armies and terrorist organizations. Furthermore, despite the extent of its territorial acquisitions, the state now had defensible borders as well. Not only did these borders follow militarily impressive topographical obstacles, such as the Suez Canal in the south and the Jordan River in the east, and not only did they incorporate militarily significant high ground, such as the Judean and Samarian highlands, but the total length of the frontiers had also been shortened considerably.

Though the post-1967 geographical status quo did not fundamentally alter the IDF's operational or tactical principles, it did affect the state's national security doctrine. Two of the three interstate wars that Israel has fought in the post-1967 period have been initiated by Arab states, while it initiated two of the three interstate wars fought in the pre-1967 period. Furthermore, the lone Israeli-initiated interstate war since the Six-Day War occurred across the only border—the one with Lebanon—where the state lacked strategic depth, where its civilian population was routinely exposed to terrorist raids or artillery bombardments. Israel's decisions to initiate—or to refrain from the initiation of—interstate war, of course, have never been made solely on the basis of military considerations; however, it does appear that the acquisition of strategic depth and defensible borders in the post-1967 period has curbed, to a certain extent, its propensity to engage in preventive or preemptive interstate war.

Israel's fundamental approach to insurgency, on the other hand, does not seem to have changed drastically as a result of the Six-Day War. Retaliation as a means to influence adversarial conduct has remained a central tool for countering this type of warfare in the post-1967 period. And Israel's approach to the more recent phenomenon of asymmetrical war against heavily armed terrorist organizations has essentially been an outgrowth of its approach to lower-level insurgency.

The primary contribution of the post-1967 borders to Israeli national security is that they have insulated the state against a catastrophic reversal during interstate war by allowing it to trade space for time. In the 1973 Yom Kippur War, the depth and defensibility provided by the Sinai and Golan "buffer zones" gave the IDF the room and time that it needed to recover from its early surprise and setbacks. Because the IDF's regular forces were able to relinquish some territory, particularly in the Sinai, instead of having to make a "life-or-death" stand along the frontiers themselves, these forces were able to wage an effective mobile defense in the north and south until ample reserve units could be mobilized and deployed to the fronts. These blocking battles stabilized the situation on both fronts, and later permitted the IDF to launch punishing counterattacks, bringing Israel victory in the war.

But control of the Sinai, Gaza, Judea, Samaria, and the Golan has also created problems for Israel. For starters, its capture of the Sinai and Golan served as powerful incentives for Egypt to initiate the 1969–70 War of Attrition and Egypt and Syria to initiate the Yom Kippur War. Moreover, mass Palestinian unrest and terrorism from the late 1980s to the present, as well as terrorism and guerrilla warfare in the South Lebanon security zone from 1985 to 2000, has called into question the utility of dominating territory that is home to hostile populations.

For these reasons, since the late 1970s, Israel's national security doctrine has moved in the direction of exchanging captured territory for formal peace treaties that are accompanied by acceptable security guarantees, including international monitoring, demilitarized zones, early warning stations, bilateral security cooperation, and so forth. Thus, Israel returned the Sinai to Egypt in the early 1980s as part of such a peace agreement and gave up large chunks of Judea, Samaria, and Gaza to the Palestinians as part of the peace process.

Its readiness to surrender additional captured territory in the future—namely, most of the remainder of Judea and Samaria and all of the Golan—is currently less assured, however, in light of its recent unhappy experiences following its unilateral withdrawals from South Lebanon in 2000 and Gaza in 2005. Rather than promote peace between Israel and its neighbors, these withdrawals had the opposite effect of emboldening Hizbullah and Hamas, respectively, into ratcheting up their terrorist campaigns against it, eventually triggering two asymmetrical wars, the 2006 Second Lebanon War and the 2008–9 Gaza War (or Operation Cast Lead).[3]

Nevertheless, in a future war, whether interstate or asymmetrical in nature, Israel is unlikely to pursue territorial objectives, at least in the form of long-term occupation. Instead, the IDF is much more likely to concentrate upon the massive destruction of opposition armies or terrorist organizations as a more effective approach to ensuring Israel's national security interests.

Allocation of Manpower

Upon its establishment in 1948, Israel had a population of 600,000–650,000 people. Collectively, the neighboring Arab states had a population that numbered into the many millions. From a military perspective, this extreme demographic imbalance, which even mass Jewish immigration could not redress, meant that the Arab states would be able to maintain sizable professional armies, while Israel would not be able to do so, for an attempt to maintain a large professional army would undermine the state's economic and social development.

In the wake of the War of Independence, therefore, Israeli defense planners decided to overcome this demographic obstacle by transforming the IDF into a militialike army. During peacetime—that is, in the absence of interstate war—the IDF would consist of a limited number of full-time professional soldiers, supplemented by a much larger pool of conscripts fulfilling their mandatory military service. These professional and conscript soldiers would be joined by a third group of soldiers, reservists, each of whom would be liable for one to several months of annual military duty after completion of mandatory service. Indeed, Israelis once fondly proclaimed themselves to be a nation of soldiers on leave for eleven months of the year. The idea behind this militialike structure, which survives to the present, has always been to keep the minimum number of soldiers in uniform during peacetime so as not to disrupt the state's economic and social progress.

The soldiers of the peacetime IDF have had two fundamental functions. First, they have been tasked with maintaining Israel's day-to-day, or current, security. Responsibility for day-to-day security has meant, in practice, engaging in border skirmishing and counterinsurgency warfare against state and nonstate adversaries. Second, they have had to prepare the IDF for interstate and asymmetrical war in order to ensure Israel's long-term, or basic, security. To achieve this objective, they have had to make sure that reserve units, which have always constituted the bulk of the IDF's warfighting potential, could be quickly and smoothly mobilized and deployed for battle. Readiness for interstate and asymmetrical war has entailed such tasks as maintaining an efficient mobilization system, training conscripts and reservists, keeping equipment in good order, and updating operational and tactical plans.

By and large, a militialike IDF has served Israeli national security quite well. Not only has it done an admirable job of defending the state in both wars and counterinsurgency campaigns, but it has also done so without causing long-term economic and social disruption. Nevertheless, this elegant solution to Israel's manpower problem has carried with it a military and diplomatic price tag. Militarily speaking, the IDF experienced a near disaster at the outset of the Yom Kippur War, because its regular forces were too small to stop the attacking Egyptian and Syrian armies at the borders. Diplomatically speaking, once mobilized for war, the IDF must either be unleashed or demobilized in short order. Israel's economy simply cannot sustain an indefinite mobilization, waiting for the often slow wheels of diplomacy to turn. Israel, in other words, has never had the luxury of time in a crisis situation.

Though Israel remains committed to a militialike IDF, signs of change in this regard have been in the air since the early 1990s. Israeli defense planners have voiced the opinion that the IDF ought to become a "slimmer and smarter" organization. The precise meaning of this phrase with respect to manpower requirements has never been made clear, but it seems to indicate a desire to rely more on professional soldiers and less on conscripts and reservists in the decades ahead.

Three major reasons account for the preference for a more professional army. Two are internal to the IDF; the third is external to it. First, as a consequence of both natural growth and mass immigration, Israel's Jewish population has surpassed the six million mark. The state, according to defense planners, now has a surfeit of military manpower, which suggests to them that it should be able to make do without universal conscription in the decades ahead. Indeed, some defense planners already profess no longer to be interested in trying to integrate problematic groups within Israeli society, especially ultra-Orthodox Jews, into the IDF.

Second, as warfare has become an increasingly high-technology affair, it has become increasingly difficult for part-time soldiers to maintain and operate state-of-the-art equipment. Although they retain significant reservist elements, several branches of the IDF, particularly the air force, navy, and military intelligence, have long relied principally on professional soldiers, precisely because of the ultrasophisticated hardware and software with which they fight. Likewise, the IDF's special operations units are more professional than in the past, reflecting the more demanding and politically delicate role that they now play in Israel's defense. The trend toward an increasingly technology-oriented force structure and organizational culture, in short, should accelerate the process of professionalization within the IDF.

Third, the evolving nature of Israeli society has combined with these changes. While it would be wrong to argue that Israeli society has sunk into a morass of "post-Zionist" self-indulgence and aimlessness, many contemporary Israelis have not been quite as ready as their parents to set aside their own self-interests and personal aspirations on behalf of the state's welfare. During the 1990s, this attitude manifested itself in an increasing unwillingness to serve in the IDF on the part of influential segments of Israeli youth. Though this trend among youth has reversed itself to a large extent since the outbreak of the second intifada (or the so-called al-Aqsa Intifada) in 2000, the IDF continues to move toward becoming a more professional army that relies heavily on volunteers.

Nevertheless, it would be a mistake to exaggerate the rate at which the IDF is shedding its militialike force structure and organizational culture. Unquestionably, the IDF will retain and rely upon a large cadre of reservists for the foreseeable future—it would need them in case of an interstate war. Moreover, the long-term trend toward individual self-fulfillment notwithstanding, most Israeli youth continue to see military service as an important rite of passage into Israeli society. Social pressure alone, therefore, would indicate that the idea of mass conscription is not currently in danger of being swept away. Still, as high-technology "force multipliers," including advanced electronic systems and precision-guided munitions (PGMs), proliferate in the Israeli arsenal, the IDF will probably become somewhat more selective about whom it recruits into its ranks.

Quality versus Quantity

Historically, the Arab–Israeli conflict has been characterized by an imbalance of military resources, at least in the realm of interstate war. Israel has had—and will continue to have—fewer soldiers and arms than its adversaries. To address the problem of "the few against the many," the IDF has consistently sought to achieve qualitative superiority with regard to both manpower and arms.

Israel's manpower pool has always been qualitatively superior to the one possessed by its adversaries. Fully aware of this advantage since the establishment of the state, the IDF has sought to cultivate its manpower asset in several ways. First, the IDF has long been known for its very realistic and rigorous training methods. The training given to air force pilots, to cite just one example, has been judged to be more demanding than the training given to pilots of any other air force. Second, the IDF has traditionally placed great emphasis on the selection of its officers. The meticulousness of the selection process, as well as the painstaking training regimen, is probably unsurpassed anywhere else in the world. Third, the IDF has carefully crafted its operational and tactical plans to maximize its manpower advantage.

Arms superiority, to the contrary, is a more recent phenomenon. The ultrasophisticated arms with which today's IDF is equipped frequently obscure the fact that, before the Six-Day War, Israeli weapons were generally not superior to—and were quite often inferior to—those in the hands of its adversaries. While the Arabs received rather up-to-date Soviet arms, Israel usually had to make do with secondhand Western weapons. Only in the quality of the arms possessed by its air force, armored corps, and intelligence corps could the IDF's arsenal be said to match those of the Arab states in qualitative terms.

The IDF achieved technological superiority in the air only after the Six-Day War, when the air force began to phase out its French aircraft in favor of American aircraft. Similarly, the IDF achieved technological superiority on the water only after the Six-Day War, when the navy incorporated the then novel fast missile boat, equipped with an indigenously designed ship-to-ship missile, into its order of battle. In the arena of land warfare, technological superiority would only be achieved in the wake of the Yom Kippur War, largely through local production of arms.

The result of Israel's persistent quest to achieve and maintain qualitative superiority has been readily evident on the battlefield. Despite suffering reverses in interstate wars, asymmetrical wars, and counterinsurgency campaigns, the IDF has never been bested by any Arab army or terrorist organization. It has been the undisputed battlefield victor in every interstate war, except for the War of Attrition, which ended in a stalemate along the Suez Canal. The IDF has also performed well in asymmetrical wars and counterinsurgency campaigns, even when it has not delivered knock-out blows to Arab terrorist organizations.

Today, the IDF remains absolutely committed to the idea of maintaining its "qualitative edge" over its adversaries in manpower and arms. This emphasis on quality, however, should not obscure the fact that its attitude toward quantity changed after the Yom Kippur War.

The IDF's traumatic experience in that war, especially during the first few days of fighting, when it incurred substantial losses in men and machines, convinced it that "quantity has a quality all its own." Over the past four decades, the IDF has grown significantly in size, to the point where its arsenal now contains thousands of tanks and hundreds of fighter-bombers.[4] These figures make its arsenal among the largest in the world. Nevertheless, the commitment to a slimmer and smarter IDF should lead to a gradual reduction in the quantity of arms moving forward.

Modes of Warfare

It may seem paradoxical that Israel, a state that has never deliberately sought to expand its territory at the expense of its Arab neighbors, has been strongly committed to maneuver (or mobile) warfare, at least insofar as concerns interstate war.[5] But the IDF's embrace of this type of warfare at the operational and tactical levels has made perfect sense for a state in Israel's position.[6]

Not only has Israel sought to wage interstate wars on Arab territory, because of its historical lack of strategic depth and defensible borders, but it has also sought to wage short wars. Its preference for short interstate wars, like its preference for wars on Arab territory, is not hard to fathom. Short wars, needless to say, cause less economic disruption than long wars. Because its economy has been especially sensitive to the dislocating effects of interstate war, Israel has had a powerful incentive to terminate wars as quickly as possible. Moreover, the Jewish people's tragic past, as well as Israel's relatively limited population (by world standards), has furnished an equally powerful incentive to end interstate wars quickly so as to keep their human costs down. Finally, Israel has concluded that terminating wars sooner rather than later reduces the prospect of foreign diplomatic and military intervention on behalf of its adversaries.

Not only has maneuver warfare offered an elegant solution to the state's territorial, economic, human, and diplomatic predicaments, but it has also played to the IDF's strength vis-à-vis its opponents' armies. This type of warfare, after all, puts a premium on quality. Based as it is on rapid movement, maneuver warfare favors better-trained, better-motivated, and better-commanded forces. Numerical superiority, on the other hand, has much less of an impact on the outcome of maneuver warfare than it does on the outcome of attrition warfare.

The IDF's actual battlefield experience in interstate wars has repeatedly reinforced its commitment to maneuver warfare. During the last stage of the War of Independence, the IDF routed the Egyptian army, driving it out of the Negev, in a maneuver warfare campaign. In its early years, therefore, the IDF built itself around mechanized infantry forces of the kind that had thrashed the Egyptian army. In the 1956 Sinai Campaign, during which Israeli forces again routed the Egyptian army, capturing the whole of the Sinai in just a few days, the IDF's armored corps and air force played conspicuously impressive roles. Thus, after the war, maneuver warfare in the IDF became synonymous with the primacy of the tank and the aircraft.

The spectacular victories of its armored corps and air force in the Six-Day War simply reinforced the IDF's commitment to maneuver warfare at the operational and tactical levels. Israel's acquisition of strategic depth and defensible borders did little to temper the IDF's resolute focus on this type of warfare. Nor did the reverses suffered by its armored and air forces at the hands of Arab anti-tank and anti-aircraft weapons in the opening days of the Yom Kippur War undermine the IDF's devotion to maneuver warfare.

To the present day, the IDF continues to advocate maneuver warfare; however, it has modified its operational and tactical principles since the Yom Kippur War. One of the more significant changes involves the shift to a more balanced mix of forces—that is, giving previously neglected branches of the IDF, such as infantry and artillery, a more prominent role in its operational and tactical plans. The IDF, in other words, has reverted to a more conventional and inclusive approach to combined arms warfare with regard to interstate war. Another significant change involves a considerably greater reliance on firepower to accomplish objectives than in the past. Signs of this new emphasis on firepower became unmistakable by the outbreak of the 1982 Lebanon War. But only in the 1990s, as it digested the lessons of the 1991 Gulf War, did the IDF acknowledge, albeit rather quietly, that mobility alone may no longer represent an ideal solution on the modern Middle Eastern battlefield.

Given the "saturated" nature of this battlefield, where room for maneuver has been severely degraded by the vast numbers of weapons in Middle Eastern arsenals, the next interstate war—should there be one—might well see the IDF defer maneuver warfare until it has undertaken an intensive preparatory bombing campaign, using all manner of air-, sea-, and ground-launched PGMs against its adversaries' military forces. The IDF might well seek to weaken its opponents to such an extent that a maneuver campaign could then be carried out at low cost to itself. If the Israeli home front were to come under intense missile/rocket bombardment at the outset of a war, however, the IDF might well opt for maneuver warfare sooner rather than later in an effort to end the attacks by seizing launching areas.

Israel's approach to asymmetrical war and insurgency, on the other hand, has been quite different. Attrition warfare has been the dominant model in the former. In the Second Lebanon War, the IDF employed massive firepower from the air and land in an attempt to degrade Hizbullah's rocket arsenal and to chew up its other military assets. Throughout most of the fighting, the IDF's maneuver warfare effort was confined to small and rapid ground raids into Hizbullah strongholds close to Israel's northern border. Only in the last days of the war did the IDF engage more aggressively in maneuver warfare by sending its infantry, armored, and artillery forces deeper into southern Lebanon. Similarly, the Gaza War falls within the realm of attrition warfare. After the air force battered Hamas targets for a week, the IDF entered Gaza and spent a further two weeks moving at a snail's pace in order to pulverize what remained of that organization's military assets.

With respect to counterinsurgency campaigns, the story is much the same. Despite engaging in incursions into neighboring states that might in themselves be considered to fall under the heading of maneuver warfare, the IDF has mainly employed attrition warfare to parry terrorist organizations. From the campaign to counter the Palestinian fedayeen (or "self-sacrificers") of the early 1950s to the campaign to counter the second intifada of the early twenty-first century, the IDF has sought to grind down terrorist organizations into eventual submission by inflicting unsustainable casualties on them over time.

Israel has employed attrition warfare in asymmetrical wars and counterinsurgency campaigns for three basic reasons. First, these types of conflict have not represented an unmanageable threat to the state. However destructive of life and property, neither asymmetrical war nor insurgency has ever posed an existential threat to Israel. Second, they have not put undue stress on its internal resources. Israel's economy and society, in fact, have proven able to weather them with relative ease. And, third, neither asymmetrical war nor insurgency has ever been amenable to swift battlefield victories in the same way as interstate war, not least of all because Israel is a liberal democracy that observes moral and legal restraints in its fight against terrorist organizations, in effect fighting with "one hand tied behind its back." For these reasons, then, Israel has opted to contain the threats posed by asymmetrical war and insurgency through attrition warfare that is intended to restore the status quo ante.

Proliferating Threats

Israel has been plagued by the threat of both interstate war and insurgency since its establishment. Asymmetrical war has become a problem in the twenty-first century, with the rise of heavily armed terrorist organizations able to fire thousands of rockets into Israeli territory from southern Lebanon and Gaza. The state has also faced the threat of warfare waged with weapons of mass destruction since Egypt's use of poison gas in Yemen during the early 1960s. Nevertheless, the relative impact of these threats on Israel's national security doctrine has shifted significantly over time. The most useful distinction to make in this connection is between the pre- and post-Yom Kippur War periods.

In the pre-Yom Kippur War period, Israel's national security doctrine concentrated overwhelmingly on the threat of interstate war. Israeli defense planners, to be sure, realized that border skirmishes with Arab states and infiltration by Palestinian terrorists constituted a chronic threat, one that the IDF had to be prepared to counter. Given Egypt's supply of poison gas, they also took the threat posed by weapons of mass destruction seriously enough to mount a sabotage campaign against that state's effort to build ballistic missiles to deliver it. Still, contrary to the threat posed by interstate war, Israeli defense planners did not view these threats as representing genuine dangers to the survival of Israel.

The allocation of the state's defense resources in the pre-Yom Kippur War period illustrates this fact. Only a small proportion of Israel's resources were invested in perimeter defenses—frontier outposts, border patrols, anti-terrorist units, and minefields—intended to counter infiltrators. And only a small proportion of its resources were invested in the development of nuclear arms, as a weapon of "last resort," and anti-chemical weapons gear. The lion's share of Israel's defense resources went into the IDF's preparation for interstate wars.

In the post-Yom Kippur War period, Israeli defense planners have continued to view interstate war as a serious threat to the state's survival. Since that war, however, insurgency, asymmetrical war, and weapons of mass destruction have come to be seen as much more potent threats to the state's welfare than in the past. The upgraded status of insurgency grew out of the first intifada and the rise of Hizbullah. The recent attention to asymmetrical war is accounted for by the proliferation of medium- and long-range rockets in the hands of Hizbullah and Hamas. And the upgraded status of warfare with weapons of mass destruction stems from the proliferation of these weapons, as well as ballistic missiles, throughout the Middle East since the 1980s.

The rise of these nonconventional warfare threats has been reflected in Israel's defense resource allocations since the Yom Kippur War. While the state continues to invest heavily in preparations for interstate war, more and more assets have been devoted to other threats, particularly from the 1980s onward. These assets include the formation of special operations units specifically intended to counter terrorists, the development of anti-rocket systems designed to shoot down short-, medium-, and long-range rockets, and an entire range of active and passive defense systems intended to counter weapons of mass destruction.

To summarize, the military context shaped by these five variables—strategic depth and defensible borders, allocation of manpower, quality versus quantity, modes of warfare, and proliferating threats—helps to bring into sharper focus how and why Israel has developed and employed its airpower, the subject of the rest of this monograph. Suffice it here to say that these variables have contributed significantly to Israel's long-standing emphasis on airpower.

Introduction (Part II) Israeli Airpower and the Arab–Israeli Conflict

In late July 1970, Israel Air Force (IAF) interceptors engaged in a large-scale air battle with their Egyptian counterparts in the airspace above the Suez Canal.[1] In a short but intense battle reminiscent of a Second World War dogfight, with aircraft maneuvering against each other at close range and high speed, IAF interceptors shot down five Egyptian aircraft at no cost to themselves. Other than the fact that Soviet pilots happened to be flying the Egyptian aircraft on this particular day, this melee did not represent an exceptional event for either the IAF or the State of Israel.[2]

Indeed, over the course of its rather brief lifespan, Israel has been involved in a seemingly endless string of armed confrontations. It has fought no less than six interstate wars: the 1947–49 War of Independence, the 1956 Sinai Campaign, the 1967 Six-Day War, the 1969–70 War of Attrition, the 1973 Yom Kippur War, and the 1982 Lebanon War. Israel has also engaged in two asymmetrical wars with nonstate terrorist organizations: the 2006 Second Lebanon War with Hizbullah and the 2008–9 Gaza War (or Operation Cast Lead) with Hamas. Lastly, it has engaged in chronic counterinsurgency campaigns—for example, the first and second Palestinian intifadas—during the past six plus decades, as well as numerous special operations.

Consequently, why a state in Israel's position would seek to create a highly potent air force is not too difficult to comprehend. Israel has never possessed abundant strategic depth, even at its maximal territorial limits in the immediate aftermath of the Six-Day War, certainly in comparison to the vast majority of other states in the international system. Nor has it possessed easily defensible borders throughout most of its existence. A chronic inferiority in numbers of fighting men and machines, at least in relation to its state opponents, has been another significant hindrance to Israel's ability to protect its national interests. Under these circumstances, Israeli defense planners concluded that the IAF could compensate substantially for geographical, manpower, and matériel deficiencies by operating as a "force multiplier" that not only would be able to offer vital protection against massive attack on Israel's hinterland, but would also be able to provide considerable support to Israel Defense Forces (IDF) land forces in their offensive and defensive efforts. The IAF, in other words, would constitute a very flexible instrument for parrying the various threats faced by Israel.

Airpower and Interstate Wars

Israel has not always had a highly potent air force, however.[3] In light of the origin of the IDF in prestate militias that had no conception of how to employ airpower, of the inability of the Jewish community under the British Mandate for Palestine to build even the most rudimentary infrastructure on which to construct an air force, and of the desperate struggle for survival in which the new state found itself upon its establishment, the IAF contribution to the victory in the War of Independence proved to be relatively modest. By the end of the war, to be sure, it had acquired some reasonably modern Second World War-era aircraft, with which it had achieved air superiority over the neighboring Arab air forces. And it had undertaken sporadic close air support (CAS) and interdiction sorties on behalf of land forces. It had even carried out a few "strategic" attacks on Arab capitals. However, the IDF's triumph in the war stemmed overwhelmingly from the ability of its infantry, both foot and mechanized, to outfight and outmaneuver its Arab opposition. The most important contribution of airpower to the Israeli victory rested in its ability to fly supplies from overseas to Israel and then on to besieged strongholds and mobile columns.

Although the IDF would continue to favor mechanized infantry as the decisive combat element in warfare until the 1956 Sinai Campaign, Israeli defense planners nevertheless recognized the value of airpower in the wake of the War of Independence. Thus, from 1949–56, they provided enough resources for the IAF to develop into a first-rate combat branch, though one still quite limited in striking power. Due to resource constraints, the IAF's leadership quickly decided to emphasize quality over quantity. Rigorous selection and training processes turned out a small cadre of highly qualified air and ground crews. Limitations on the number of aircraft that could be obtained by the air force spurred the acquisition of multirole fighter-bombers rather than more specialized machines.[4] Moreover, the IAF developed an operational plan consonant with the IDF's focus on offensive maneuver warfare. At the outset of hostilities, it would destroy the opponent's airpower on the ground, thereby achieving air superiority early in the contest, which would then leave it free to lend a hand, through CAS and interdiction sorties, to the land forces in their thrust(s) into Arab territory.

In the event, the IAF did not receive permission to carry out its prewar operational plan. Still, the Sinai Campaign constituted a turning point in the IAF's history. Its very efficient performance in the war—dropping parachute troops on target deep behind Egyptian lines during the opening day of the conflict, achieving air superiority over the Sinai battlefield by repeatedly besting Egyptian aircraft in air battles, carrying out CAS and interdiction sorties on behalf of the advancing land forces, and resupplying those units as they moved swiftly across the Sinai peninsula—did not go unnoticed by Israeli defense planners. In the aftermath of the Sinai Campaign, therefore, the IAF—along with the armored corps, which had also performed very efficiently in the fighting—moved from the margins to the very center of the IDF's operational doctrine.

The war had vindicated the IDF's decision to build itself primarily to wage offensive maneuver warfare; however, it had also revealed that mechanized infantry should no longer be at the core of its operational doctrine. Instead, this doctrine would henceforth be based on the apparent supremacy of the tank and the airplane. This decision resulted in a major expansion of the IAF from 1956–67. The pool of air and ground crews grew in both quality and quantity, as did the inventory of aircraft, ordnance, ancillary equipment, and air base facilities. Of equal importance, the air force refined its own operational doctrine during this interwar period. It committed itself even more firmly to the concept of a first strike against the opposition's airfields in order to destroy its aircraft on the ground. Surviving aircraft would then be swamped and destroyed in air-to-air combat by the IAF's fleet of interceptors. The IDF's land forces would be asked to make do with minimal air support while the contest for air superiority was underway. Only after air superiority had been achieved would the IAF allocate its squadrons to the CAS and interdiction roles.

The IDF's stunning victory in the Six-Day War vindicated the armor–air combination in the eyes of Israeli defense planners. The IAF, after all, thoroughly destroyed the Egyptian, Jordanian, and Syrian air forces on the ground at the outset of hostilities, furnishing Israel with air superiority over the battlefields of Sinai, Judea, Samaria, and the

Golan. It then turned its attention to flying hundreds of interdiction and CAS sorties on each front. And it also supplied critical logistical support to the IDF's land forces, especially in the Sinai. No wonder, asserted Israeli defense planners, that the IDF's armored spearheads easily crushed the Egyptian, Jordanian, and Syrian armies in six days, leading to the capture of Sinai, Judea, Samaria, and the Golan. Airpower, in the Israeli estimation, had proven itself to be king of the battlefield.

The outcome of the Six-Day War led to another round of expansion for the IAF, to the point that, by the end of the 1960s, it had unmistakably become the most important branch of the IDF, receiving a very large share of the defense budget. The quantity and quality of its aircraft grew by leaps and bounds in the half-decade following the war, particularly as it rather quickly transitioned from French to American aircraft.[5] The far superior range, ordnance capacity, and electronic systems of American aircraft meant that the IAF could now deliver much greater quantities of more advanced munitions, including first-generation precision guided munitions (PGMs), over far greater ranges than in the past, and with better accuracy to boot. In terms of striking power, the American A-4 Skyhawk and F-4 Phantom represented an enormous upgrade over the French Mirage III, Vautour, Super Mystère, Mystère IV, and Ouragan. And, once more, the IAF deepened and improved its pool of air and ground crews. Doctrinal changes accompanied these changes in force structure. Specifically, the air force acknowledged that it would most likely no longer be able to destroy an opponent's airpower on the ground at the beginning of hostilities now that Egypt and Syria had hardened their airfields and built integrated air defense systems (IADSs) to protect both their air and land forces. The IAF concluded that its main challenge in future rounds of fighting would be knocking out those IADSs in order to achieve air superiority over the battlefield.

The IAF had to implement these changes in force structure and doctrine in the midst of the War of Attrition, a war in which Israel relied on airpower to inflict a decisive defeat on its opponent. Israel, though, did not decisively defeat Egypt in the war; rather, hostilities ended in a stalemate along the Suez Canal. The IAF had proved itself more than able to best the Egyptian air force in air combat, had proved itself more than able to launch strategic attacks deep inside Egypt's hinterland, and had proved itself able to land hard blows against Egyptian land forces, including IADS positions, along the Suez Canal; however, the IAF had not proved itself able to achieve air superiority in the canal zone by overcoming the Egyptian IADS.

Israeli defense planners deemed the War of Attrition to be an aberration. They believed that the next war would revert to the pattern of the Sinai Campaign and the Six-Day War—that is, it would be a war of maneuver, not attrition. Hence, the IDF did not alter its basic military doctrine in the three years that separated the War of Attrition and the Yom Kippur War—it continued to premise its operational plans on the supremacy of armor and airpower. For its part, the IAF emerged from the War of Attrition firmly convinced that it must be given a free hand to attack the opponent's IADS(s) at the outset of the next war, without the distraction of flying CAS and interdiction sorties in support of land forces. To that end, the IAF developed and refined operational plans to launch massive sustained attacks against both the Egyptian and Syrian IADSs upon the outbreak of hostilities.

Alas, the Yom Kippur War caught Israel by surprise; therefore, the IAF never got a chance to implement its prewar operational plans. Instead, it had to engage in large-scale CAS and interdiction strikes on behalf of hard-pressed land forces in the Sinai and on the Golan, which were heavily outnumbered by attacking Egyptian and Syrian forces, respectively, during the opening phase of the war. Accordingly, though the IAF ultimately made a significant contribution to Israel's undisputed victory in the war, it did not have the impact on the fighting that the IDF had anticipated before the start of the war. While the IAF protected Israel's hinterland, shot down hundreds of Egyptian and Syrian aircraft in air combat, restricted the overall pace and scope of the Arab war effort, helped to stem the initial Syrian offensive on the Golan, and carried out effective strategic attacks on Arab infrastructure targets, it did not obtain clear air superiority over the battlefields, at least until the final phase of the war on the Sinai front, because it was not afforded the opportunity to concentrate exclusively on overcoming the Egyptian and Syrian IADSs. Moreover, the IAF suffered heavy losses to those IADSs in the process of executing CAS and interdiction sorties that, collectively speaking, did not strongly influence the course of events in the ground war.

In the wake of the Yom Kippur War, the IDF modified its military doctrine somewhat. While it remained committed to offensive maneuver warfare as its primary warfighting modus operandi, the primacy of the armor–air combination gave way to a more balanced combined arms approach that upgraded the role of infantry, artillery, engineers, and other combat branches previously neglected under the pre-Yom Kippur War doctrine. This doctrinal alteration did not have a major direct impact on the IAF, as its share of the defense budget remained very high and as its emphasis on the destruction of the opponent's IADS(s) at the outset of hostilities reflected its pre-Yom Kippur War operational plans. To meet this goal, the IAF acquired the most sophisticated aircraft (e.g., the F-15 Eagle and F-16 Fighting Falcon), ordnance (e.g., the Tadmit television-guided missile and GBU-15 television-guided glide bomb), and electronic systems in the world from both the United States and local industry, and integrated them under the aegis of a highly sophisticated battle management system. And, in what would become a sort of mini precursor to the Revolution in Military Affairs (RMA) of the late twentieth and early twenty-first centuries, in which state-of-the-art technologies have been woven together seamlessly in a sensor-to-shooter network, it pioneered the use of unmanned aerial vehicles (UAVs) to furnish real-time intelligence to both air and land forces.

In contrast to the Yom Kippur War, Israel took the initiative in the Lebanon War, its last interstate war to date, so the IAF got the opportunity to execute its prewar operational plan with regard to the Syrian IADS in the Bekaa. In an enormous coordinated strike, the IAF essentially obliterated the entire IADS there without loss to itself. When Syrian air force interceptors rose to confront the IAF, they were destroyed in droves, with 80–100 being lost in a series of air battles in which not a single Israeli aircraft went down. Nevertheless, even though the IAF obtained air superiority over the battlefield, its CAS and interdiction sorties in support of the IDF's land forces did not heavily influence the ground fighting against either Syria or the Palestine Liberation Organization (PLO), in large measure due to the mountainous and urban terrain of Lebanon, which placed both topographical and moral constraints on IAF attacks.

Airpower and Sub-interstate War Conflicts

The IAF has a long history of involvement in armed confrontations other than interstate war, too. The first large-scale employment of airpower in such a conflict took place in the mid-1960s, when the IAF helped to put a stop to Syrian efforts to divert the Jordan River's waters from flowing into Israel by knocking out earth-moving equipment on the Golan. The IAF also engaged in occasional skirmishes with Egyptian, Jordanian, and Syrian aircraft along Israel's borders throughout the 1950s and 1960s.

The intensive use of airpower in armed confrontations other than interstate war, however, dates from the immediate post-Six-Day War years, when the IAF became a key player in the IDF's counterinsurgency campaign against the PLO. IAF aircraft repeatedly struck PLO targets in Jordan and, later, Lebanon, and air force helicopters routinely transported infantry forces in "hot pursuit" of guerrilla bands. The IAF's struggle against PLO forces would continue—and even intensify in the 1970s and early 1980s—until that organization was coerced into abandoning Lebanese territory as part of the agreement to end the Lebanon War. From the mid-1980s until 2000, the IAF's counterinsurgency campaign in Lebanon focused principally on destroying Hizbullah targets in southern Lebanon, especially during two major IDF escalations of the fighting, Operation Accountability in 1993 and Operation Grapes of Wrath in 1996, which were intended to suppress rocket fire against northern Israeli towns and villages.

With the outbreak of the second intifada (or the so-called al-Aqsa Intifada), the employment of airpower in counterinsurgency duties has become even more common. Since the turn of the century, helicopter gunships and, perhaps, UAVs have routinely been used to carry out "targeted attacks" against Hamas, Islamic Jihad, and Fatah leadership targets in Judea, Samaria, and Gaza.[6] The IAF has employed UAVs for round-the-clock surveillance and intelligence gathering in these areas, and helicopter gunships have been used to support armored and infantry formations involved in counterinsurgency duties. Both helicopter gunships and aircraft have attacked stationary targets, such as arms depots, rocket launching sites, training facilities, and so forth.

Undoubtedly, though, the most intensive employment of airpower in armed confrontations other than interstate war has occurred during the two asymmetrical wars fought by Israel over the past decade against Hizbullah and Hamas, respectively. In the Second Lebanon War, Israel relied upon airpower to deliver a knockout blow to Hizbullah. Despite some extremely impressive achievements during the fighting—for instance, smashing Hizbullah's entire long-range rocket capability early on in the conflict—the massive application of airpower, as in the War of Attrition, did not result in a decisive victory for Israel, which proved unable to suppress completely rocket fire against its northern population centers. In light of its overreliance on airpower in the war against Hizbullah, the IDF opted for a more balanced approach in its operation against Hamas in Gaza. The initial use of airpower during Operation Cast Lead stunned Hamas, and it inflicted heavy losses in men and matériel on the organization before the IDF's land forces entered the area. It was the steady advance of these forces into the heart of Hamas-controlled Gaza, however, that eventually brought the organization to its knees.

Finally, the IAF has engaged in many special operations over the decades. Some of these operations have been executed in cooperation with IDF infantry units, like the anti-terrorist hostage-rescue mission at Entebbe airport, Uganda, in 1976. Other operations—such as the destruction of an Iraqi nuclear reactor in 1981, the destruction of PLO headquarters facilities in Tunisia in 1985, and the destruction of a Syrian nuclear facility in 2007—were entirely the work of the IAF.

With regard to operational plans for the employment of airpower in armed confrontations other than interstate war, the IAF has essentially functioned in an improvisational fashion. In contrast to its operational plans for interstate war, formal doctrinal principles, such as the necessity of destroying an opponent's IADS at the outset of hostilities, have not historically guided its conduct in this realm of warfare. Since the turn of the century, though, the IAF has begun to ruminate more systematically about the use of airpower in sub-interstate-war scenarios, as this type of conflict presently appears to be the norm for Israel. This thinking has already been reflected, for example, in the development of a highly sophisticated airborne web that integrates aircraft, helicopter gunships, and UAVs in a rapid-response network.

Not with Aircraft Alone

The IAF's contribution to Israeli national security, however, has not been based solely on the employment of airborne vehicles in interstate wars and other types of conflict. Beginning in the early 1970s, the air force has continually augmented and upgraded its own IADS. During the Yom Kippur War, ground-based air defenses shot down a large number of Arab aircraft while defending IDF land forces against aerial attacks. The rise of ballistic missiles and rockets—and the concomitant decline of aircraft—as a threat to Israel over the past two decades has led to a significant reorientation of the IAF's IADS. Today, the Arrow anti-ballistic missile interceptor system and the Iron Dome short-range rocket interceptor system offer a substantial amount of protection not only to Israel's civilian and industrial centers, but also to strategic targets, like air bases and research facilities. David's Sling (also known as Magic Wand)—optimized to intercept medium-range rockets, cruise missiles, and aircraft—will come online in 2013–14 to fill the gap between the Arrow and Iron Dome systems in an effort to seal Israel's airspace against hostile fire.

Furthermore, the IAF is in charge of Israel's growing fleet of reconnaissance satellites. From the first experimental vehicles launched into space in the late 1980s and early 1990s, the IAF now has at its disposal at least four highly sophisticated reconnaissance satellites. These platforms, which carry optical and radar systems, complement the intelligence furnished by the IAF's other reconnaissance assets in the form of aircraft and UAVs. The precise capabilities of these satellites, as well as the nature of the intelligence data produced by them, not surprisingly, are closely held secrets; nevertheless, it is certain that Israel's eyes in space keep a close watch on hostile states, like Syria and Iran, and hostile terrorist organizations, like Hizbullah and Hamas.

And neither the IAF's aircraft, IADS, nor reconnaissance satellites, of course, could function effectively without an extensive and integrated network of command and control (C^2) facilities, bases, and technical schools to coordinate and support them.

Humanitarian Operations

The IAF has contributed to Israel's national security in a more pacific fashion as well. Two major airlifts in the 1980s and 1990s brought most of the Ethiopian Jewish community to Israel, thus fulfilling a cardinal principle of the state's foreign policy—the necessity to bring home imperiled Jewish communities around the world. And the IAF's participation in numerous overseas humanitarian relief operations from the mid-1980s to the present has generated a certain amount of diplomatic goodwill toward Israel among states that have often been cool toward it over the years. The actual humanitarian work has been done by the IDF's medical corps and home-front search and rescue (SAR) teams; however, without the IAF's substantial airlift capabilities, Israeli medical and rescue personnel would not have had the opportunities to do their work.

Definitions of IAF Aircraft Roles

Before proceeding any further, it is first necessary to define in general terms the main roles of the IAF's aircraft in warfare (as these roles will be examined in the next few chapters). The IAF has traditionally had four main combat roles—air superiority, CAS, interdiction, and strategic attack—as well as four ancillary, "noncombat" roles—troop transport, casualty evacuation, logistical support, and reconnaissance.[7]

An air force that has achieved air superiority in a conflict is one that essentially controls the skies over the theater of operations. Such an air force is able to carry out all of its assigned roles—for example, CAS and interdiction—secure in the knowledge that its aircraft face only minimal opposition from an opponent's air force and IADS. Such an air force is also able to deny its opponent's air force the capability to fulfill these same roles by ensuring that enemy aircraft cannot survive in sufficient numbers to be effective.

Air supremacy is an extreme form of air superiority. An air force that has obtained air supremacy over the theater of operations is one that has gained virtually unchallenged command of the skies such that its own aircraft can complete their assignments at almost no risk to themselves and such that its opponent's aircraft stand almost no chance of survival, let alone of executing their assignments.

Both CAS and interdiction involve attacks on an opponent's land forces (or other targets) situated on the battlefield. The distinction between these roles is the proximity of the attacks to friendly land forces. CAS involves attacks against an opponent's land forces (or other targets) in very close proximity to friendly forces. It requires precise coordination between air and land forces to assure the avoidance of "fratricide." Interdiction, in contrast, involves attacks against an opponent's land forces (or other targets), which, while on the battlefield, are not in very close proximity to friendly forces. This type of sortie does not require the same high level of air–land coordination, as the prospect of fratricide is not a consideration. Interdiction strikes may also be carried out against targets that are away from the battlefield itself—for example, rear area supply convoys and transportation infrastructure.

Strategic attack sorties are directed against military or nonmilitary (but war-related) targets—for instance, leadership personnel; central headquarters; missile or rocket launchers; munitions depots; petroleum, oil, and lubricants (POL) storage dumps; and port facilities—that are sometimes located on the battlefield itself, but often are situated in an opponent's hinterland. These attacks are often characterized, in other words, by the "deep penetration" of an opponent's airspace, and they are intended to serve some larger purpose than merely obtaining a tactical advantage on the battlefield—for example, they may be intended to undermine an opponent's ability to wage war by "decapitating" its leadership or by destroying its war-related infrastructure, or they may be intended to defend one's own homeland from serious threats, such as those posed by massed missile or rocket fire.

The four ancillary roles can be defined more succinctly. Troop transport concerns the movement of land forces to, from, around, and behind the battlefield. Casualty evacuation involves the removal of the wounded (and, sometimes, the dead) from the battlefield. Logistical support concerns the supply of land forces with consumables, such as vehicles, munitions, POL, food, and water. And, finally, reconnaissance involves data collection with respect to the location and strength of the opponent's forces and targets, on or off of the battlefield, and with respect to damage assessment of forces and targets engaged previously.

1

Airpower and Maneuver Warfare: The Israel Air Force in the 1967 and 1973 Wars

The 1967 Six-Day War and the 1973 Yom Kippur War began quite differently for the Israel Air Force (IAF). During the opening day of the Six-Day War, on the basis of a meticulously planned and exhaustively rehearsed operational plan, which it refined and updated in the three weeks prior to the outbreak of hostilities, the IAF assumed the initiative against the air forces of Egypt, Jordan, and Syria, taking them by surprise on the ground. The first few days of the Yom Kippur War, to the contrary, found the IAF scrambling to respond to an Egyptian–Syrian surprise attack against the State of Israel. Though the IAF had carefully planned and thoroughly rehearsed operational plans, in this instance aimed primarily at Arab integrated air defense systems (IADSs) rather than at Arab air forces, the diplomatic and military circumstances prior to the onset of hostilities prevented it from seizing the initiative.

The conditions prevailing at the outset of each war determined the IAF's performance in the opening phases of these conflicts. In the Six-Day War, the IAF essentially annihilated the combined Arab air forces of Egypt, Jordan, and Syria in a few hours by launching several waves of highly coordinated air base attacks. The virtually complete elimination of Arab airpower—and the concomitant early achievement of air superiority over the battlefields—then left it free to concentrate on other roles in support of the Israeli war effort. During the Yom Kippur War, on the other hand, the IAF did not have the opportunity to secure air superiority over the battlefields early in the conflict, because the Israel Defense Forces (IDF) high command insisted that it fly close air support (CAS) and interdiction sorties on behalf of hard-pressed land forces. With sustained operations against Arab IADSs not an option—

and with these systems subjected only to intermittent and hesitant IAF strikes—they exacted a significant toll on Israeli aircraft at modest cost to themselves during the first few days of the war.

The immense victory scored by the IAF against the Arab air forces during the Six-Day War, followed by the equally impressive triumph registered by IDF land forces against the Egyptian, Jordanian, and Syrian armies, has created the distinct impression that airpower played an overwhelming role in that victory. Contrariwise, the early setback suffered by the IAF at the hands of Arab IADSs during the Yom Kippur War, coupled with the much longer period of time that it took the IDF to defeat the Egyptian and Syrian armies in comparison to the Six-Day War (almost three weeks versus less than one), has created the distinct impression that airpower did not make a major contribution to the Israeli victory.

Admittedly, these impressions seem valid upon first reflection. With Arab air forces quickly eliminated as a threat both to Israel proper and to IDF land forces, the IAF participated extensively in the ground battles on the Sinai (Egyptian), Judean and Samarian (Jordanian), and Golan (Syrian) fronts during the Six-Day War, inflicting substantial damage on Arab armies on particular occasions. The long lines of burned out Egyptian army vehicles in and around the Sinai passes serve as mute testimony to this fact. The IAF's inability in the Yom Kippur War to neutralize the Egyptian and Syrian IADSs, especially in the early phase of the conflict, hindered its capacity to support the IDF's land forces. That the Egyptian army could move tens of thousands of troops and thousands of vehicles across the Suez Canal during the first days of the war with only minor losses caused by the IAF serves as eloquent testimony to this fact.

A careful examination of the relative contribution of Israeli airpower to the 1967 and 1973 war efforts, however, reveals a far less clear-cut, much more complex tapestry. Though its significant contribution to the IDF's victory in the Six-Day War is undeniable, the IAF did not win the war for Israel. Nor can the IAF's performance in the Yom Kippur War be deemed the principal reason why it took the IDF so long to defeat its Arab opponents. Its initial problems with Arab IADSs notwithstanding, the IAF played a rather substantial role in the IDF's triumph.

Perhaps the best means with which to gauge the actual contribution of Israeli airpower to the Israeli war efforts in the 1967 and 1973 wars is to compare the IAF's performance in each of its four main combat roles—air superiority, CAS, interdiction, and strategic attack—as well as in each of its four ancillary, "noncombat" roles—troop transport, casualty evacuation, logistical support, and reconnaissance—across both wars. But it is first necessary to describe in broad strokes the IAF's accomplishments (or lack thereof) in each war, so that its relative performance can be judged accordingly.

A Summary of IAF Activity in the 1967 War

Following three weeks of fruitless diplomacy to resolve an Arab–Israeli crisis that had erupted as a consequence of border tensions between Israel and Syria, the Israeli government gave the IDF a green light to commence operations against the Egyptian armed forces in early June 1967. The IAF immediately launched Operation Focus, a full-scale preemptive strike on the Egyptian air force that had been planned and practiced for years.[1]

Attacking at an unusual time—well after dawn—and from an unexpected direction—largely from west to east—the IAF caught the Egyptian air force completely unprepared for battle. In a multiwave assault that employed nearly its entire inventory, the IAF continually struck 18 air bases in the Sinai and in Egypt proper throughout the first day of the fighting according to a predetermined ranking of targets that gave priority to the destruction of long-range bombers and frontline interceptors. Moreover, the air bases themselves were thoroughly worked over in repeated bombing and strafing runs, and the IAF also struck some of Egypt's radar stations.[2]

Operation Focus devastated the Egyptian air force. Israeli figures record the destruction of approximately 300 aircraft, including the entire long-range bomber fleet and most of the frontline interceptor fleet. Many of the air bases were rendered more or less inoperative, because IAF aircraft had cratered their runways and demolished their facilities, and the Egyptian air force possessed only a rudimentary repair capability. The Egyptian air force, in sum, had been reduced to a mere shadow of its former self in terms of aircraft, air bases, and command and control (C^2) infrastructure.[3]

The Syrian, Jordanian, and Iraqi air forces responded to the IAF's assault on the Egyptian air force by initiating a number of small-scale air attacks against Israel. These strikes caused no appreciable damage, but they did alert the IAF to the potential threat posed by these air forces. The IAF, therefore, embarked on a concerted campaign of air base attacks against them. It promptly and repeatedly struck eight air bases in Syria, Jordan, and Iraq as part of Operation Focus.

The results were much the same as those registered in the Egyptian portion of the operation, though on a considerably smaller scale. The IAF destroyed about half of the Syrian air force, including many of its frontline interceptors. Jordan lost almost all of its combat aircraft, and Iraq lost a number of planes. All together, the IAF destroyed about 90 Syrian, Jordanian, and Iraqi aircraft. Air bases were again pummeled to the point where they became unserviceable, with runways cratered and facilities smashed beyond easy repair.[4]

While Operation Focus exacted a considerable toll on the IAF, which lost 18 of the approximately 250 combat aircraft (including armed trainers) in its prewar inventory, it was now largely free to support IDF land forces.[5] The IAF's CAS and interdiction effort got underway in earnest on the second day of the war. Of the 2,591 CAS and interdiction sorties flown by the IAF throughout the war, a mere 268 occurred on the first day.[6] Most of the ground attack sorties throughout the war were interdiction strikes, as the IAF had neither the C^2 infrastructure nor an adequate number of the proper type of aircraft (i.e., low- and slow-flying straight-winged platforms) to engage in a high-tempo CAS effort in close coordination with advancing IDF land forces.

On the Sinai front, the IAF typically launched strikes far in the Egyptian rear, gradually working its way back toward an ever-shifting front line.[7] Most of the damage inflicted on the Egyptian army by air attack occurred during its panic-stricken retreat through the central Sinai passes toward the Suez Canal on the third and fourth days of the war. And most of the vehicles destroyed by the IAF were "soft" targets, such as trucks and jeeps, which composed the Egyptian army's logistical "tail." Relatively few tanks and other armored fighting vehicles were knocked out by Israeli airpower.[8] The IAF, in other words, did not cause heavy damage to the Egyptian army's frontline units and operational reserves, the "teeth" that engaged the IDF's land forces.

The Judean and Samarian front witnessed a similar story, even if the IAF "softened up" this front for the IDF's land forces to a greater extent at the outset of the fighting. The IAF placed much of its emphasis on interdiction strikes against logistical and infrastructure targets located in the Jordanian army's rear. While the IAF did not pin down or obliterate the army's frontline forces, it did scatter or stall at least some units from the operational reserves rushing to the front line, and it did inflict substantial damage on other units retreating back toward Jordan proper. And the IAF probably also prevented an Iraqi expeditionary force from reaching the front. While the IAF most likely had a somewhat greater impact on the ground battles in Judea and Samaria than in the Sinai, it by no means compromised the fighting ability of the Jordanian army.

The situation on the Golan front differed from those on the Sinai and Judean and Samarian fronts in that the IAF flew hundreds of interdiction sorties in the days prior to the commencement of the IDF ground assault against the Syrian army. These air attacks had a negligible effect on the heavily bunkered defensive positions on the Golan—IAF munitions of the late 1960s could not penetrate them—so the impact on the Syrian army's frontline units turned out to be principally psychological in nature. Interdiction strikes, on the other hand, did disrupt road traffic, inhibiting the capacity of the operational reserves to bolster the frontline positions. Airpower, though, again made itself felt most intensely during the retreat phase of the fighting.

A rather small air force (by major power standards) in 1967, the IAF consequently sought to maximize the number of combat aircraft in its inventory. Hence, its transport and helicopter fleets got the short end of the stick. Yet, the IAF did possess enough assets in these areas to fulfill ancillary roles. It transported paratroopers behind Egyptian lines on at least three occasions and behind Syrian lines on at least one occasion.[9] The IAF also removed wounded soldiers from the battlefields on all three fronts. Logistically speaking, fuel drops allowed a number of hard-charging IDF armored columns to continue virtually uninterrupted their advance toward the Suez Canal.[10] And, finally, the air force carried out regular reconnaissance flights, with combat aircraft serving in a dual role, for both battle-damage-assessment and target-location purposes on all fronts.

A Summary of IAF Activity in the 1973 War

Unlike the Six-Day War, when the IDF had three weeks in which to mobilize and deploy for war, the Yom Kippur War caught Israel by surprise. The IDF's land forces, particularly the reserve armored divisions that made up the bulk of its fighting power, were neither mobilized nor deployed along the Sinai and Golan fronts. The IAF, on the other hand, which has always been much less dependent on reserve manpower than the land forces, had enough advanced warning of the impending Egyptian and Syrian assault to launch a preemptive strike against their IADSs. The IAF required just a few hours to ready itself to implement Operation Challenge 4 (Sinai front) and Operation Model 5 (Golan front). Under intense American pressure, however, the Israeli government refused to sanction a preemptive strike.[11] The IDF had to absorb the first blow in this war.

The IAF, therefore, never got a genuine opportunity to implement its intricate prewar operational plans to launch massive coordinated attacks against both the Egyptian and Syrian IADSs, which it perceived to be the major obstacles to the attainment of air superiority over the Sinai and Golan battlefields. The IAF did attempt to implement Operation Challenge 4 at the start of hostilities, meeting with initial success, but the IDF high command quickly canceled the operation in order to divert the IAF's full strength to the Golan front in an effort to prevent a potential Syrian armored breakthrough into Israel proper.[12] Instead the IAF had to make do throughout the war with hastily improvised, piecemeal attacks against both IADSs at times when it had resources available for this task.

Not surprisingly, then, the results of its attacks against the Egyptian and Syrian IADSs proved to be decidedly mixed. On the Sinai front, in a series of combined operations with IDF land forces that had crossed to the Egyptian side of the Suez Canal, the IAF eventually managed to pierce the Egyptian IADS, destroying about one third of its surface-to-air missile (SAM) batteries, opening up undefended flight corridors, but only during the final phase of the war.[13] On the Golan front, in contrast, the IAF could not punch holes in the Syrian IADS, though it thinned out the number of SAM batteries stationed at the front. Emblematic of the IAF's troubles on this front, a much scaled-down and very short-lived version of Operation Model 5 executed on the second day of the war resulted in the loss of six F-4 Phantom fighter-bombers without leading to the destruction of any SAM batteries. At war's end, less than 15 percent of Syria's SAM batteries had been destroyed by the IAF.[14]

If the IAF had a rough time with the Egyptian and Syrian IADSs, which accounted for the majority of the approximately 105–110 Israeli aircraft (including helicopters) lost throughout the war, the same cannot be said of its encounter with the Egyptian and Syrian air forces.[15] It shot down approximately 275–300 Arab aircraft in air-to-air combat for the loss of perhaps 15–20 of its own.[16] Whatever the precise number of losses on each side, the IAF's Mirage IIIs, Neshers (an indigenously developed version of the Mirage), and Phantoms reigned supreme over Arab MiG-21s, MiG-19s, MiG-17s, Su-7s, and Su-20s in air battles throughout the war, even on the first day of the fighting, when they not only downed many Egyptian fighter-bombers, but also a substantial number of troop-carrying helicopters trying to drop Egyptian special operations units behind Israeli lines.

Moreover, the IAF flew hundreds of air base attack sorties during the war, mainly against Egyptian airfields, despite knowing that these sorties would destroy only a handful of aircraft on the ground and would close down only a handful of air bases for short periods of time (because Arab air forces had hardened their facilities and improved their repair capabilities after the Six-Day War debacle).[17] It flew these sorties primarily in order to keep both the Egyptian and Syrian air forces and IADSs on the defensive.

Of the more than 11,200 sorties flown by Israeli aircraft during the Yom Kippur War, approximately 7,300 (or about two thirds) were devoted to CAS or interdiction.[18] Because the IDF's land forces were not ready to counter the Egyptian–Syrian surprise attack, the IAF had to operate as "flying artillery" during the opening days of the war in order to "fill the gaps in Israeli defensive positions," especially on the Golan front.[19] Rather than begin a dedicated CAS and interdiction campaign after delivering a sharp blow to Arab IADSs, as called for in its prewar operational plans, the IAF had to conduct this campaign throughout the war, often in the face of intense anti-aircraft fire.

On the Sinai front, the IAF initially concentrated its CAS strikes around the besieged IDF strongholds on the "Bar-Lev Line" and its interdiction strikes against the Egyptian bridgeheads over the Suez Canal.[20] Though the IAF hit a large number of the bridges that spanned the canal, the Egyptians proved able to repair them very quickly. Once the IAF's focus switched to the Golan front, it severely reduced the number of CAS and interdiction strikes on the Sinai front, especially in light of its heavy losses to the Arab IADSs during the first few days of hostilities.[21] Until IDF land forces counterattacked across the Suez Canal and began to demolish SAM batteries on the Egyptian side of the canal, the IAF's CAS and interdiction effort concentrated largely on Egyptian forces operating outside of their air defense umbrella. Only in the last phase of the war, with the penetration of the system, did the IAF intensify its CAS and interdiction effort on this front, in order to support the IDF's counteroffensive into Egypt.

Out of necessity, the IAF operated according to a different set of rules on the Golan front. Because the IDF high command feared an imminent Syrian armored breakthrough into Israel proper during the early days of the war, the IAF had to engage in a high-tempo CAS and interdiction effort on this front at the outset of the war, regardless of the cost to its A-4 Skyhawk and Phantom squadrons. Only after the mobilization and deployment of the IDF's reserve armored divisions stopped the Syrian offensive and stabilized the front did the IAF's effort slacken somewhat in intensity in order to conserve aircraft. The IAF also furnished CAS and interdiction strikes in support of the IDF's counteroffensive into Syrian territory.

Like the results of its campaign against Arab IADSs, the IAF's CAS and interdiction effort yielded decidedly mixed results. Notwithstanding the tremendous growth in the firepower of the IAF between the 1967 and 1973 wars, its ability to destroy tanks and other armored fighting vehicles remained quite modest.[22] It possessed neither the required C^2 infrastructure nor the required munitions to be very effective in this regard. Still, the "shock effect" of CAS and interdiction strikes sometimes served to stall or scatter Arab frontline and operational reserve units on both fronts.

The IAF, as in the Six-Day War, had much more of an impact against thin-skinned logistical vehicles. Indeed, on the Golan front, IAF attacks on supply convoys disrupted the Syrian offensive, giving IDF land forces valuable extra time to mobilize and deploy for defense. On the Sinai front, IAF attacks on supply convoys contributed to the Egyptian Third Army's inability to prevent IDF land forces from surrounding and besieging it at the end of the war.

In contrast to the Six-Day War, the IAF conducted a series of strategic attacks in the Yom Kippur War, albeit of limited scope and duration. In response to Syrian medium-range rocket attacks against civilian and military targets in northern Israel, the IAF carried out at least one bombing raid on the Syrian capital, Damascus, causing damage to both the defense ministry complex and air force headquarters. It also pulverized POL storage and port facilities deep within Syria.

Between 1967 and 1973, the IAF upgraded its transport and helicopter fleets, thereby increasing its capabilities in the troop transport, casualty evacuation, and logistical support roles. During the Yom Kippur War, it employed helicopters to carry troops to the battlefield on a number of occasions, perhaps most notably during the last phase of the war on the Golan front, when they landed paratroopers and special operations units on Mt. Hermon in order to retake an Israeli intelligence-gathering post that had been overrun early in the war. Helicopters also evacuated large numbers of wounded soldiers from the battlefields over the course of the fighting. IAF transport aircraft not only brought munitions directly from the United States to Israel, but they also ferried supplies to the battlefields, particularly after IDF land forces captured airfields on the Egyptian side of the Suez Canal.

The IAF also upgraded its reconnaissance assets in the interwar period. During the 1973 war, it used both dual-role combat aircraft and helicopters to gather intelligence. It employed the latter mainly to monitor Egyptian and Syrian SAM batteries, while it used the former to carry out battle-damage assessments and to collect information on troop movements and rear-area targets.

A Comparative Assessment of the IAF in the 1967 and 1973 Wars

In both the 1967 and 1973 wars, the IAF achieved air supremacy over Israel proper. Very few Arab aircraft penetrated Israeli airspace in either conflict, and none of them inflicted any real damage on rear-area targets. An Egyptian attempt to attack the Israeli defense ministry complex in downtown Tel Aviv with air-to-surface missiles during the Yom Kippur War also ended in total failure.[23] The small number of medium-range rockets that Syria fired into northern Israel in this war caused more damage than Egyptian and Syrian air attacks during the conflict.

The attainment of air supremacy over Israel constitutes the most impressive contribution made by the IAF to both Israeli victories. In the Six-Day War, neither Israel's civilian populace nor its industrial assets came under air attack. Had the Arab air forces not been destroyed at the outset of the war, they might have been able to do considerable damage inside Israel's borders before the termination of hostilities. In the Yom Kippur War, air supremacy was even more important to Israel's triumph. First, Israel's civilian populace and industrial assets were again spared from death and destruction. Second, because the IDF's land forces had not been mobilized and deployed along the fronts weeks before the outbreak of hostilities, air supremacy over Israel furnished them with the crucial 48 hours of breathing space that they needed to mobilize and deploy fully to the fronts. Had Arab aircraft been able to penetrate Israeli airspace, they might have been able to disrupt the Israeli mobilization effort by attacking key staging areas, which in turn might well have changed the complexion of the war.

In both the 1967 and 1973 wars, the IAF achieved air superiority over the battlefields. In the Six-Day War, the early destruction of Arab air forces—and the consequent air superiority enjoyed by the IAF—not only allowed the air force to begin a dedicated CAS and interdiction campaign on the second day of the war essentially free of concern about interception, but it also meant that IDF land forces never came under effective air attack throughout the fighting. The few CAS and interdiction sorties flown by surviving Arab aircraft merely pricked IDF land forces; these sorties in no way impaired the IDF's offensives in the Sinai, in Judea and Samaria, or on the Golan.

The air superiority story in the Yom Kippur War is far more complex. From day one of the war, the IAF achieved air superiority vis-à-vis the Egyptian and Syrian air forces. While it is certainly true that these air forces were far more active throughout the war than they were in 1967, it is equally true that they inflicted only minor damage on IDF land forces throughout the fighting. Even on the first day—when the Egyptian and Syrian air forces launched their most intense CAS and interdiction strikes of the war—they achieved only meager results, as the IAF effectively disrupted their attacks. The impact of these air forces on the battlefields went downhill from here, and no evidence exists to support the notion that either air force inhibited IDF defensive or offensive operations later in the war. Furthermore, neither the Egyptian nor the Syrian air force was able to prevent the IAF from engaging in CAS or interdiction strikes at any stage of the conflict.

The Egyptian and Syrian IADSs, though, did seriously impair the IAF's CAS and interdiction campaigns on the Sinai and Golan fronts. Though the IAF achieved a rather tenuous air superiority vis-à-vis the Egyptian IADS by the end of the war, it nevertheless still had far from an entirely free hand over the battlefield; and it had even less success against the Syrian IADS, despite the fact that this network also had been thinned out by the end of the conflict.

Indeed, the most important accomplishment of the IAF with respect to Arab IADSs did not occur on the battlefields themselves, but rather took place in the realm of deterrence. The obsessive fear of Israeli airpower on the part of Egyptian and Syrian defense planners encouraged them to load down their armies with anti-aircraft defense units at the expense of additional armored and infantry formations. Moreover, these planners limited the scope and pace of the initial offensives on both fronts such that Arab land forces would not advance beyond the range of their slow-moving IADSs. The Egyptian penetration of the Sinai did not exceed a few miles, while the Syrians did not get much further on the Golan. The skewed force structure of Arab land forces, the limited scope of their opening offensives, and the slow pace at which they advanced at the outbreak of the war all contributed significantly to the IDF's ability to block them from moving further into the Sinai or into northern Israel itself.[24]

The attainment of air superiority over the 1967 and 1973 battlefields also constitutes a very impressive contribution to Israel's triumphs. While the IDF, because of its indisputable qualitative superiority over Arab armies, would most likely have won both wars even had the IAF not achieved air superiority, the fighting, particularly in the Yom Kippur War, would almost certainly have taken a much heavier toll on it.

The IAF's own CAS and interdiction campaigns in both wars played much smaller roles in Israel's victories.[25] In the Six-Day War, IDF land formations won the all-important "break-in" battles—that is, the first battles of its offensives, the ones that ultimately decided the outcome of the fighting—on the Sinai and the Judean and Samarian fronts with little or no direct assistance from the IAF. When given a choice to fight with airpower during the day or to fight without it at night, IDF division commanders preferred the second option.[26] Even on the Golan front, where the IAF softened up the battlefield for days before the ground assault, the IDF triumphed over the Syrian army mainly because its land forces ousted the Syrians from their stout defensive positions in a grueling slugging match. The IAF CAS and interdiction campaign, however, by compounding Arab confusion and panic and by inflicting substantial damage on soft targets, did lead to a swifter and more crushing Israeli victory in the war than would otherwise have been the case.

During the Yom Kippur War, the IAF's CAS and interdiction campaign also produced limited results. Its most notable contribution to the Israeli victory occurred on the Golan front, where it helped to stem the Syrian offensive. Nevertheless, credit for the IDF's ultimately successful blocking battles on both the Sinai and Golan fronts belongs primarily to the regular and reserve armored and infantry units that fought the Egyptian and Syrian armies to a standstill, despite being heavily outnumbered in both arenas. Likewise, credit for the IDF's successful counteroffensives in the Sinai and on the Golan again belongs mainly to the land forces. CAS and interdiction strikes hammered some Egyptian and Syrian frontline and operational reserve units that wandered outside the umbrella of their IADSs, but these strikes did not alter the course of the ground war.

A contrast is often drawn between the allegedly devastating IAF CAS and interdiction campaign in 1967 versus the allegedly anemic campaign in 1973.[27] But a close examination of the results of these campaigns—in terms of the amount of physical destruction visited upon the Egyptian and Syrian armies, though not necessarily in terms of the amount of psychological devastation inflicted on them—reveals that the difference between the two is not all that large. The IAF proved unable to destroy significant numbers of tanks and other armored vehicles in either war. While the shocking scenes of mangled convoys, so prevalent in the Six-Day War, may not have been evident in the Yom Kippur War, the IAF appears to have been equally successful at knocking out soft and rear-echelon targets in both conflicts. And aircraft losses in these campaigns were comparable when adjusted for the numbers of CAS and interdiction sorties flown in each war.[28] Whether the CAS and interdiction campaign could have made a much larger contribution to the Israeli war effort during the Yom Kippur War had the IAF been afforded the opportunity to deal first with Arab IADSs must remain an open question.[29]

The IAF's strategic attacks against Syria during the Yom Kippur War had only a marginal impact on the Israeli war effort. Though the strikes themselves caused substantial damage to infrastructure targets in the Syrian hinterland, they did not detract from Syria's overall war-making capacity, perhaps because of their restricted scope. They did, however, reinforce Israeli deterrence with respect to the home front, as the Syrians refrained from launching any further rocket attacks into northern Israel once the IAF began to hit rear-area targets. Furthermore, strategic attacks served to thin out Syria's IADS on the Golan, as the Syrian army had to move anti-aircraft defense formations to the rear in order to guard sensitive infrastructure targets. This redeployment, in turn, made it easier for IAF aircraft to fly CAS and interdiction sorties.

In terms of its ancillary roles, the IAF had about the same limited impact on the Israeli war effort in both the 1967 and 1973 wars. Troop transport to, from, and around the battlefields was confined almost exclusively to the insertion of paratroopers and special operations units behind Arab lines. With one or two exceptions in each war, these drops did not play a major part in land force operations. Casualty evacuation turned out to be a much more significant use of the IAF's airlift capability, probably saving scores of lives in each war. Likewise, logistical support in the form of IAF-delivered supplies proved important at specific points during both wars. Finally, while the IAF possessed high-quality reconnaissance aircraft, the information collected by these platforms proved to be of scant value to the IDF in both wars, because this data could not be processed and distributed to the relevant commands in real time.[30]

Conclusion

That the IAF made substantial contributions to Israel's victories in both the 1967 and 1973 wars is not in doubt. Most importantly, it ensured that Israel would not lose either war, which was—and still is—crucial to a state whose opponents contested—and still contest—its very right to exist. The IAF's contributions to Israel's victories with respect to its direct impact on the 1967 and 1973 battlefields in support of the IDF's land forces, on the other hand, were more modest in scope.

The attainment of air supremacy over Israel proper, as well as the attainment of air superiority over the battlefields, not only spared the state's civilian populace and industrial assets, but also cleared the air, so to speak, for the IDF's land forces to come to grips with their Arab counterparts. Indeed, because of the radically different set of circumstances under which the IDF began the 1967 and 1973 wars, the IAF actually played a more important part in the latter victory, in the sense that it provided the crucial time and space for the IDF's land forces to mobilize and deploy for battle. Those land forces, not the IAF, however, ultimately defeated the Arab armies in both wars.

The most fundamental lesson to be drawn from the IAF's experiences in the 1967 and 1973 conflicts is that airpower cannot be relied upon to be decisive on the conventional battlefield, whether in a maneuver or an attrition war. If more proof is necessary, one need look no further than Israel's two most inconclusive wars, the 1969–70 War of Attrition and the 2006 Second Lebanon War, the two conflicts in which it relied most heavily on airpower to accomplish its wartime objectives.

2

Airpower and Attrition Warfare: The Israel Air Force in the 1969–70, 2006, and 2008–9 Wars

Even though the Israel Defense Forces (IDF) has historically emphasized maneuver warfare, the State of Israel has been involved in a number of wars of attrition, and the Israel Air Force (IAF) played a prominent part in each one of them. The 1969–70 War of Attrition pitted Israel against Egypt in an interstate "trench-style" war fought primarily along the banks of the Suez Canal. The 2006 Second Lebanon War and the 2008–9 Gaza War (or Operation Cast Lead), to the contrary, found Israel confronting nonstate terrorist organizations, Hizbullah and Hamas, in asymmetrical wars in Lebanon and Gaza, respectively.

Israeli airpower faced different challenges in each war. In the War of Attrition, the IAF confronted an opponent equipped with an air force, an integrated air defense system (IADS), and land forces very generously supplied with artillery pieces. In the Second Lebanon War, it grappled with the best-trained, best-organized, and best-equipped terrorist organization in the world, whose military capabilities included a very sizable arsenal of long-, medium-, and short-range rockets. In the Gaza War, the air force took on a more modestly trained, organized, and equipped terrorist organization with considerably less firepower at its disposal. Nevertheless, despite these differences, it is worthwhile to compare the IAF's experience during all three wars.

As in the cases of the 1967 Six-Day War and the 1973 Yom Kippur War, perhaps the best means with which to gauge the actual contribution of Israeli airpower to the Israeli war efforts in the 1969–70, 2006, and 2008–9 wars is to compare the IAF's performance in each of its four main combat roles—air superiority, close air support (CAS), interdiction, and strategic attack—as well as in each of its four ancillary, “noncombat” roles—troop transport, casualty evacuation, logistical support, and reconnaissance—across all three wars. But it is first necessary to describe in broad strokes the IAF’s accomplishments (or lack thereof) in each war, so that the air force’s relative performance can be judged accordingly.

A Summary of IAF Activity in the 1969–70 War

Though the War of Attrition got underway in earnest in spring 1969, when Egypt formally repudiated the cease-fire agreement that ended the Six-Day War, hostilities had actually begun soon after the guns had fallen silent in the previous round of fighting. In preparation for the “liberation” of the Sinai, Egyptian forces sporadically harassed the IDF, mainly with artillery barrages and special forces assaults against positions on the east bank of the Suez Canal. Israel, naturally, responded in kind with artillery barrages and special forces assaults of its own. Israeli airpower largely stayed out of hostilities during this phase of the war, because Israel sought to limit the scope and intensity of the fighting and because the Egyptian air force, in the process of rebuilding itself after its devastating defeat in the Six-Day War, was not ready for a major confrontation with the IAF.

Nevertheless, in this tit-for-tat skirmishing, the IAF had two tasks: first, to maintain control of the airspace over the combat zone by shooting down any Egyptian aircraft that ventured there and, second, to ferry special operations forces to targets deep in the Egyptian hinterland. From the summer of 1967 to the spring of 1969, the IAF destroyed approximately 10 Egyptian aircraft in air battles.[1] Control of the airspace over the combat zone, however, offered no protection against Egyptian artillery barrages, which took a toll in IDF personnel. In the autumn of 1968, in an attempt to compel Egypt to put an end to these barrages, therefore, IAF helicopters transported special operations forces to attack three targets—a power station and two bridges—far in the Egyptian rear.[2] The targets themselves suffered only minor damage, but the message that Israel could—and would—strike Egypt’s undefended rear in response to continued shelling struck a nerve. The artillery barrages came to an immediate halt.

But the deterrent effect of this operation wore off after a few months. In the spring of 1969, Egypt renewed its artillery barrages, this time around on a considerably larger and much more sustained basis. Neither IDF artillery fire against Egyptian army positions on the west bank of the Suez Canal nor additional special operations raids in the

Egyptian heartland sufficed to silence the Egyptian guns.[3] Hence, in the summer, Israel decided to "escalate [the war] for the sake of de-escalation," in the words of then IDF Chief of Staff Chaim Bar-Lev.[4]

Operationally speaking, Israel now committed the IAF to a full-scale interdiction campaign against the Egyptian army's artillery positions, as well as its emerging IADS, along the Suez Canal. For six months, throughout the summer and autumn of 1969, the IAF, which functioned as "flying artillery" to compensate for the paucity of land-based firepower in the IDF, systematically pounded Egyptian artillery batteries, surface-to-air missile (SAM) batteries, anti-aircraft artillery (AAA) batteries, radar stations, and other targets. It also engaged in air battles with a more aggressive Egyptian air force, and it continued to ferry special operations forces on raids deep inside Egypt.[5] The air force intensified its logistical support to land forces in the Sinai by flying in much-needed supplies, and it evacuated increasing numbers of wounded IDF personnel to rear-area hospitals as well.

By the winter of 1969, the IAF had essentially destroyed the canal-side IADS, had shot down approximately another 35 Egyptian aircraft in air battles, and had inflicted heavy personnel losses on Egyptian forces.[6] Still, these achievements proved insufficient to stop the incessant Egyptian artillery barrages, even if the IAF's "counterbattery" fire muted their impact on IDF positions along the canal. Every time the IAF knocked out one Egyptian gun, it seemed, another popped up to take its place; hence, Israel once more decided to escalate the War of Attrition in an effort to end hostilities on its terms.

The method chosen for this second round of escalation involved strategic attacks on military targets in close proximity to the Egyptian capital and other major cities. In addition to keeping up the pressure along the Suez Canal and participating in special forces raids, the IAF initiated a "deep-penetration" bombing campaign in early 1970.[7] For four months in the winter and spring, F-4 Phantoms conducted approximately 40 air strikes—118 sorties in all—against SAM batteries, radar stations, training bases, supply depots, and the like.[8] Regardless of their effectiveness in destroying their targets, however, these long-range raids did not produce the desired result.

Indeed, not only did the raids fail to compel Egypt to end the War of Attrition, but they also provided a pretext for direct Soviet intervention in the fighting. Partly in response to this Soviet move, and partly in response to American political pressure, Israel called an immediate halt to the IAF's deep-penetration campaign. Nevertheless, throughout the spring and summer of 1970, thousands of Soviet military advisors, technicians, and pilots poured into Egypt, accompanied by more advanced SAM batteries and aircraft than the Egyptians possessed at the time.[9] And it was not long before these troops became actively enmeshed in the fighting.

With very considerable assistance from the Soviet Union, the Egyptian army eventually built a robust IADS in the Suez Canal zone, replacing the one destroyed in the previous year. The IAF furiously contested the construction of this more comprehensive and sophisticated IADS, inflicting heavy losses on Egyptian and Soviet men and machines in the process, and it continued to reign supreme in air-to-air combat, shooting down approximately another 45 aircraft during this stage of the war, including five piloted by Soviet flyers; however, the IAF gradually lost control of the airspace above the Suez Canal in the final months of the war, losing approximately 20 aircraft in its ultimately unsuccessful effort to stop the rolling advance of the Egyptian–Soviet IADS toward the canal.[10] By the middle of the summer of 1970, both Israel and Egypt were worn out by their bruising encounter in the War of Attrition, allowing the superpowers finally to broker a cease-fire agreement.

A Summary of IAF Activity in the 2006 War

Following the outbreak of the second Palestinian intifada in 2000, Hizbullah provided regular assistance to Palestinian terrorist organizations, such as Hamas, in carrying out homicide bombings and other attacks against Israeli targets. And it occasionally heated up the Israel–Lebanon frontier by sending squads of terrorist operatives to infiltrate northern Israel. Not surprisingly, then, when one such squad, under cover of a rocket barrage against northern Israeli villages, ambushed an IDF border patrol in the summer of 2006, killing several soldiers and capturing two others (who subsequently died in Hizbullah's hands), Israel decided that it had had enough of these provocations, and gave the IDF the go ahead to launch a major campaign against the organization.

The IAF took center stage in the IDF's operational plan. For most of the 34 days of the war, it shouldered the burden of Israel's war effort. By one account, the IAF flew over 15,000 sorties during the war, including 10,000 ground attack sorties (2,000 of them by helicopter gunships) and more than 2,500 reconnaissance and transport sorties.[11] In the course of its air campaign, the IAF struck over 7,000 targets: headquarters facilities, training and logistical bases, rocket launchers, vehicles, bridges, airport runways, and so forth.[12] IDF land forces, in contrast, engaged mainly in rather small-scale raids on Hizbullah positions close to the border, at least until the final days of the war, when they undertook a more sweeping assault, up to the banks of the Litani River; the artillery branch, however, did fire tens of thousands of shells and rockets at Hizbullah targets throughout the war.

The IAF engaged in strategic attacks against two distinct types of target: command and control (C^2) facilities/personnel and rocket launchers.[13] With regard to the former, these attacks caused tremendous physical destruction to Hizbullah facilities in southern Beirut and in the Bekaa, but they did not significantly impede the organization's ability to communicate with its units.[14] Even the destruction of an Iranian-built C^2 facility for coordinating rocket salvos against Israel appears to have had no real impact on Hizbullah's rocket-launching effort.[15] Nor did IAF strikes kill many, if any, high-ranking commanders of the organization.

With regard to the latter set of targets, the IAF chalked up some quite impressive achievements. In the course of hostilities, the IDF (i.e., primarily the IAF) knocked out at least 125 rocket launchers, including most of Hizbullah's long- and medium-range launchers, and may have smashed as many as 250 others.[16] Indeed, early on in the war, the IAF eliminated 90–95 percent of the organization's long-range rocket launchers, taking a potent threat to the Israeli heartland off of the table.[17] None of these rockets hit Israeli population centers.

Hizbullah's more numerous medium-range rockets inflicted some damage on the Israeli home front, particularly in the port city of Haifa; however, by the end of the war, the IAF also had little trouble knocking out Hizbullah's medium-range rocket launchers. On the last day of the fighting, for example, the air force destroyed seven medium-range rocket launchers.[18] The IAF's highly advanced "24/7 sensor-to-shooter" network meant that Israeli aircraft often destroyed these launchers within two minutes of detection, before some multibarreled platforms could fire more than one or two rounds.[19] All in all, Hizbullah's medium-range rocket launchers did not play a central part in the hostilities.

The IAF, to the contrary, had no effective answer to Hizbullah's short-range rockets. Of the approximately 4,000 rockets that landed inside Israel during the war, including over 200 on the last day of hostilities, the vast majority fit into this category.[20] The simple, low-technology design of these short-range rockets rendered them essentially impervious to the IAF's high-technology response. Easily transported and concealed, and with a minimal launch signature, these rockets could not be struck by the air force in real time. Hizbullah terrorist operatives would simply "shoot and scoot" before the IAF could engage in counterbattery fire. Moreover, Hizbullah's exploitation of southern Lebanon's population as human shields—rockets were frequently located within or in close proximity to homes, schools, mosques, and medical facilities—meant that the IAF could not conduct "saturation bombing" of known launch areas, as Israel sought to minimize collateral damage to that population.[21]

In addition to strategic raids, the IAF mounted vigorous interdiction attacks during the war. Israel intended to impede Hizbullah's ability to resupply its units in southern Lebanon, as well as to impede its ability to receive provisions from abroad, via interdiction strikes against roads, bridges, airport runways, warehouses, and fuel dumps. The stockpiling of provisions in southern Lebanon prior to the war rendered the air force's effort in this respect largely moot. Hizbullah's units there did not suffer from crippling shortages of munitions, fuel, or other basic provisions. The IAF's effort to block Iran and Syria from resupplying Hizbullah with arms via the Beirut airport and overland routes, on the other hand, almost certainly bore some fruit.[22]

The IAF engaged in vigorous CAS attacks in support of IDF land forces, too. Helicopter gunships like the AH-1 Cobra and AH-64 Apache, with their high-velocity cannons and anti-armor missiles, proved particularly suitable in this role, often striking targets only a few meters away from IDF infantry and armor.[23] Hizbullah's entire military infrastructure in southern Lebanon was shattered during the fighting, and some of the credit for this outcome must be given to the IAF's CAS strikes. Still, the level of air–land coordination during the fighting satisfied neither the air force nor the land forces.

Finally, the IAF made a considerable contribution to the Israeli war effort in its ancillary roles. First, it transported special operations forces behind Hizbullah lines on at least 20 separate occasions, including a spectacular raid on a facility in the Bekaa that netted several prominent prisoners and killed about 20 terrorist operatives.[24] It also transported infantry behind Hizbullah lines during the IDF's large-scale ground offensive in the last days of the war. Second, the IAF medevaced a considerable number of wounded IDF personnel back to Israel, often pulling these men out of "hot" landing zones.[25] And, third, the IAF's unmanned aerial vehicles (UAVs), as well as its manned platforms, carried out a very large number of reconnaissance and battle damage assessment sorties throughout the war, many of which were critically important to the effort to destroy Hizbullah's rocket arsenal.

A Summary of IAF Activity in the 2008–9 War

Not long after the outbreak of the second intifada (or the so-called al-Aqsa Intifada), Hamas (and other terrorist organizations, such as Islamic Jihad) began a sustained campaign of rocket and mortar fire against Israeli towns and villages around Gaza. This campaign grew in scope and intensity after Israel's unilateral withdrawal from Gaza in the autumn of 2005, as Hamas began to smuggle (through tunnels under the Egyptian border) and stockpile substantial quantities of more advanced, longer-range rockets. By the winter of 2008, major Israeli population centers located considerable distances from the Gaza border had come under regular rocket bombardment; therefore, in order to curtail the rocket and mortar campaign, Israel directed the IDF to undertake a major operation against Hamas (and its partners).

Operation Cast Lead, a combined air–land assault, lasted for approximately three weeks.[26] Unlike the War of Attrition and the Second Lebanon War, Israel did not call upon the IAF this time around to shoulder the burden of hostilities largely on its own. The operation did begin with a week of intensive air strikes against Hamas facilities—rocket launch sites, arms factories and storage areas, headquarters facilities, training bases, administrative centers, smuggling tunnels, and so on—but the final two weeks witnessed an extensive ground maneuver by the IDF's land forces, with the IAF serving in a supportive capacity.

Similar to its role in the Second Lebanon War, the IAF conducted strategic attacks against two distinct types of target: C^2 facilities/personnel and rocket launchers.[27] A massive opening strike pulverized Hamas headquarters facilities, administrative centers, and training bases, not to mention other sorts of targets, throughout Gaza, collectively causing hundreds of casualties among its ranks.[28] Furthermore, "targeted attacks" killed or incapacitated a significant number of senior- and middle-ranking Hamas commanders during the course of the war. From the first day of hostilities, therefore, the organization more or less lost control over its forces in the field, which tended to act in a haphazard and disjointed manner.

In respect of its anti-rocket launcher effort, the IAF registered notable achievements. In the first days of the war, Hamas (and allied) forces were able to fire approximately 60 rockets per day into Israel. By the end of the war, this figure had dropped to approximately 20 per day.[29] Moreover, the rockets that hit Israel caused few casualties and little damage. Nor did they induce much of the civilian population to flee the area, as had occurred in northern Israel during the Second

Lebanon War. Not all of the credit for the ineffective rocket attacks against Israel, of course, belongs to the IAF, as IDF land forces played a substantial part in the effort to destroy the launchers in the final two weeks of fighting; however, the IAF's pinpoint attacks during the opening phase of hostilities, as well as its support of the land forces later on, certainly contributed heavily to the outcome.

The air force also engaged in intensive interdiction and CAS strikes against Hamas targets. With respect to the former, the IAF devoted special attention to smuggling tunnels and arms storage facilities. Hamas could not bring in meaningful supplies of arms from outside Gaza during the hostilities as a result of these strikes, and much of its local stockpile went up in flames before it could be employed on the battlefield. With regard to the latter, the IAF functioned in very close cooperation with IDF land forces, even integrating forward air controllers into specific units.[30] The air force used its assets—especially helicopter gunships—not only to provide "covering fire" for advancing land forces, but also to detonate improvised explosive devices and other booby traps ahead of these forces.[31] The small number of Israeli casualties in the war—the IDF suffered only 10 dead—testify to the efficacy of its CAS strikes.

Finally, as in the Second Lebanon War, the IAF made a substantial contribution to the Israeli war effort in its ancillary roles. Not only did it transport troops to the combat zone, but it also evacuated the wounded from the battlefield. Its greatest contribution in this sphere, however, lies in the incessant, round-the-clock reconnaissance activity carried out particularly by its UAV squadrons. Constant surveillance of the battlefield permitted near instantaneous precision attacks against rocket launchers, command personnel, squads of terrorist operatives, and other targets of opportunity, and provided IDF officers with an accurate picture of unfolding events, including post-strike damage assessments.

A Comparative Assessment of the IAF in the 1969–70, 2006, and 2008–9 Wars

In both asymmetrical wars, not surprisingly, the IAF maintained air supremacy over both the Israeli home front and the battlefield. During the Gaza War, Hamas possessed no SAM or AAA batteries; therefore, it had no way to engage the IAF's aircraft, helicopter gunships, or UAVs, which flew over targets at will and suffered no losses in the fighting. Nor did Hamas command any air assets with which to strike

Israel. During the Second Lebanon War, Hizbullah possessed at least some anti-aircraft capabilities in the form of hand-held SAMs and AAA batteries, but it had nothing like an IADS.[32] Furthermore, the IAF's extensive use of electronic countermeasures degraded even this rudimentary anti-aircraft capability. Though the IAF did suffer one loss to anti-aircraft fire, a transport helicopter in the final days of the war, its aircraft, helicopter gunships, and UAVs flew over the battlefield without limit throughout the course of hostilities. Unlike Hamas, Hizbullah made two attempts to penetrate Israel's air defenses with attack drones. Of the four UAVs employed in these attempts, two succumbed to technical malfunctions and two others were shot down by IAF aircraft before they could strike Israeli targets.[33]

The air superiority story in the War of Attrition is more complex. The war took place far from Israel proper, so the Israeli home front was not involved in the fighting. Over the Suez Canal battle zone, the IAF maintained air superiority vis-à-vis the Egyptian air force throughout the war. It shot down scores of Egyptian aircraft for the loss of just a handful of its own; and, at no time did IDF positions along the canal come under sustained, effective air attacks from the Egyptian air force. Initially, the IAF achieved air superiority vis-à-vis the Egyptian IADS, allowing its aircraft to roam freely not only over the Suez Canal zone, but also throughout the Egyptian interior; however, once the Soviet Union inserted itself into the fighting, the IAF quickly curtailed its deep-penetration raids, thereby forfeiting air superiority in the Egyptian hinterland. The IAF lost air superiority over the Suez Canal zone within months thereafter, as the Egyptians rebuilt their canal-side IADS with extensive Soviet support. The IAF ended the War of Attrition with control of the Sinai airspace, but not over any Egyptian-held territory.

In contrast to the Six-Day War and the Yom Kippur War, strategic attacks were central to the IAF's war effort in each of the three wars of attrition. In the Gaza War, the air force's strategic attacks, especially the opening aerial blitz, shredded Hamas' C^2 infrastructure, decapitated a portion of its command element, and disrupted its rocket-launching campaign, blows from which the organization did not recover during the fighting. Consequently, both the organization's rocket attacks against the Israeli home front and its defense against the IDF's ground incursion turned out to be very ineffective. The IAF's strategic attacks, in other words, had a major impact on Israel's decisive victory in the war.

The air force's strategic attacks in the Second Lebanon War had a less salubrious effect on the outcome of hostilities. While IAF attacks caused severe damage to Hizbullah's C^2 infrastructure, they did not noticeably disrupt the organization's means of communicating with its forces in the field, nor did they incapacitate its command element. Hizbullah's ability to direct its forces in the field remained robust throughout the war. Furthermore, while IAF attacks crippled Hizbullah's long-range rocket arsenal at the outset of the war, as well as curbed the destructiveness of its medium-range rocket arsenal during the fighting, thereby reducing the magnitude of the damage to the Israeli home front, they could not counter the organization's short-range rocket arsenal, so northern Israel continued to be heavily bombarded to the last day of the war. The IAF's strategic attacks, in short, simply did not undermine Hizbullah's warfighting capabilities.

The IAF's strategic attacks in the War of Attrition generated the most disappointing results of all from the Israeli perspective. Despite their tactical excellence, the air force's deep-penetration raids did not subvert the Egyptian regime.[34] Nor did they sap Egypt's will or capability to fight on. Indeed, they probably stiffened Egypt's resolve to continue the war. Even more damaging to Israel, they gave the Soviet Union a convenient pretext for direct intervention in the war, an intervention that cost the IAF air superiority over the Egyptian interior and along the Suez Canal. The deep-penetration raids, in sum, triggered a process that ended in a setback for the IAF and Israel.

Like its strategic attacks, the IAF's interdiction and CAS efforts bore the most fruit in the Gaza War. Constant interdiction and CAS sorties from aircraft, helicopter gunships, and, perhaps, drones not only thoroughly obstructed Hamas' ability to resupply its forces in the field with arms, but also made it very difficult for its terrorist operatives to confront IDF land forces. IAF aircraft and helicopter gunships, under the guidance of forward air controllers and land commanders, systematically demolished the organization's prepared defensive positions, including those located in dense urban environments, often before these positions even had an opportunity to engage IDF land forces. The IAF commander at the time of the war had this to say:

> We planned well in advance for the Gaza operation, with [air force] guys spending time in ground commanders' war rooms and their guys flying with us in Cobras and Apaches. We developed a common language, and officers even recognized each other's voices on the radio.
>
> A key lesson from the 2006 war was the need for improved jointness and support for ground forces, and this is something we were able to implement in [Operation Cast Lead]. In [this

> war], we let brigade commanders on the ground have almost full control over air assets, both visual elements and attack helicopters. And based on the feedback we got from ground forces—and the decisive achievements of the operation itself—the outcome was very good.[35]

The efficient use of airpower explains to a significant extent why Hamas (and its allies) lost 70–80 terrorist operatives for every IDF soldier killed in the war.[36]

The IAF's interdiction and CAS strikes against Hizbullah in the Second Lebanon War had much less of a negative impact on the organization's fortunes than the air force's interdiction and CAS strikes during Operation Cast Lead had on Hamas' fortunes. In the six years following the Israeli withdrawal from southern Lebanon in spring 2000, Hizbullah had built a large-scale, sophisticated, and dispersed military infrastructure throughout the area. When war erupted in summer 2006, its field units had ample stockpiles of munitions, fuel, and other consumables at hand; they did not require resupply during hostilities. Though the IAF's interdiction strikes against warehouses, roads, bridges, airport runways, and the like limited the amount of war matériel reaching Hizbullah from Iran and Syria during the war, it did not impede the organization's capability to fight in the short term. If the war had gone on much longer, of course, then the IAF's interdiction strikes might have made a difference to the outcome. The air force's CAS strikes—while they certainly inflicted casualties on Hizbullah terrorist operatives, kept down the IDF's own personnel losses, and destroyed defensive positions throughout southern Lebanon—did not cause the organization's forces to abandon the battlefield. IDF land forces had to overcome Hizbullah terrorist operatives in many tough firefights.

The IAF did not engage in many, if any, CAS strikes during the War of Attrition, as IDF land forces, with the exception of occasional special operations forays, did not cross over to the Egyptian side of the Suez Canal. The air force did, however, mount an extensive interdiction effort throughout the war against Egyptian positions along the Suez Canal. These strikes, which destroyed very large numbers of artillery pieces, supply depots, C^2 facilities, observation points, and the like, helped at times to reduce the amount of death and destruction visited upon IDF land forces on Israel's side of the canal. Nevertheless, Egypt continued to strike IDF positions with artillery fire throughout the war. The IDF eventually had to construct what became known as the "Bar-Lev Line," after IDF Chief of Staff Chaim Bar-Lev, a series of heavily bunkered strongpoints along the Suez Canal, partly in recognition of the fact that the IAF could not prevent Egyptian artillery fire. The outcome of the IAF's interdiction effort, like the outcome of the war itself, proved to be inconclusive.

In terms of its ancillary roles, the IAF had a limited impact on the Israeli war effort in each of the three wars. It medevaced casualties from the battlefields to hospitals in all three wars, almost certainly saving lives in the process. It brought supplies to the front in all three wars, even airdropping them directly to front-line forces in the Second Lebanon War. It transported troops to the battlefields in all three wars, and far behind the front lines on special operations in two of them. But it is perhaps in the realm of reconnaissance that the IAF had its greatest impact, especially during Israel's two asymmetrical wars. The real-time, round-the-clock intelligence information provided particularly by the air force's UAV assets allowed both it and IDF land forces to mount devastating precision attacks against Hizbullah and Hamas targets, particularly rocket launchers in the act of firing their loads against the Israeli home front.

Conclusion

Though the IAF performed very well at the tactical level in each of the three wars of attrition, two of them ended rather inconclusively for Israel. In the War of Attrition, Israeli airpower prevented the "liberation" of the Sinai, Egypt's self-declared reason for initiating hostilities, and inflicted heavy losses in men and machines on Egyptian military forces; however, the IAF did not bludgeon Egypt into submission on Israel's terms, and it lost air superiority over the Suez Canal zone before the summer 1970 cease-fire agreement. In the Second Lebanon War, Israeli airpower inflicted heavy losses in men and machines on Hizbullah, and it substantially degraded that organization's ability to inflict damage on the Israeli home front through massed rocket fire by the time a cease-fire agreement went into effect; but, once again, the IAF did not deliver a knockout to Israel's opponent. In the Gaza War, Hamas suffered a crushing defeat; however, in this war, in contrast to the previous two wars of attrition, Israel did not rely on the IAF to win the conflict primarily on its own, as IDF land forces took a significant part in the fighting.

3

Airpower, Counterinsurgency, Special Operations, and Humanitarian Operations: The Israel Air Force between Arab–Israeli Wars

The State of Israel has waged counterinsurgency warfare of one sort or another throughout most of its history. In the pre-1967 Six-Day War era, the Israel Air Force (IAF) did not participate to any significant extent in this type of warfare. Whether the state called upon the Israel Defense Forces (IDF) to defend the frontiers against incursions by terrorist operatives or to attack their bases in neighboring Arab states, land forces shouldered the lion's share of the burden. In the post-Six-Day War era, in contrast, the air force has become a major actor in Israel's counterinsurgency efforts, both within and beyond its frontiers.

Israel has also engaged in special operations against its Arab opponents throughout its history. In the pre-Six-Day War era, with the notable exception of the acquisition of a Soviet MiG-21 from Iraq in the mid-1960s, the first time this type of aircraft had been obtained by a Western power, the IAF did not participate to any significant extent in these operations, either. The IDF's land forces once more shouldered the lion's share of the burden in this type of warfare. Since the Six-Day War, though, the IAF has engaged in numerous special operations, sometimes in concert with land forces and sometimes on its own.

Airpower, of course, has historically been used on a smaller scale in counterinsurgency and special operations warfare than in high-intensity maneuver and attrition warfare. Nevertheless, it has fulfilled the same kinds of roles in the former as in the latter; therefore, the IAF's activities in the realms of counterinsurgency and special operations warfare are also best described and assessed in light of the air force's traditional combat and "noncombat" roles.

Additionally, since the late 1970s, the IAF has been involved in a string of humanitarian operations around the world. Two of these operations—dubbed Moses and Solomon—involved the evacuation of much of the Ethiopian Jewish community to Israel via a pair of large-scale airlifts, one conducted covertly and one conducted overtly. The IAF has also assisted states faced with natural or manmade catastrophes, from Mexico in 1985 to Turkey in 2011, acting in a crucial logistical capacity in disaster relief.

A Summary and Assessment of IAF Activity in Counterinsurgency Warfare

In the aftermath of its performance in the Six-Day War, the IAF's status within the IDF soared to new heights. Consequently, it was called upon to shoulder more of the burden of Israel's day-to-day security than in the past. In respect of counterinsurgency warfare, beginning in the late 1960s, the air force teamed up with land forces to parry the tremendous upsurge in Palestine Liberation Organization (PLO) attacks against Israel. The IAF's part in this counterinsurgency campaign involved interdiction strikes against PLO bases in Jordan and close air support (CAS) strikes on behalf of land forces raiding those bases. A notable example of interdiction and CAS strikes occurred during the large-scale armored–infantry incursion against the major PLO base at Karameh.[1] Furthermore, air force helicopters routinely transported infantry squads in "hot pursuit" of terrorist operatives who had crossed the Jordan River into Israeli-held territory in Judea and Samaria, as well as special forces units on raids deep inside Arab territory.

Effective air–land coordination resulted in quite substantial personnel losses among PLO terrorist operatives by 1970.[2] Indeed, the unsustainable rate of casualties convinced the organization essentially to forgo its campaign of direct attacks against Israeli territory, at least for awhile, in favor of a terror campaign on Jewish and Israeli targets abroad in the form of bombings, airline hijackings, assassinations, and so on. The IAF's interdiction and CAS strikes—along with the capability to transport land forces to the battlefield and deep behind the front lines—certainly contributed in a crucial way to the defeat of the PLO in the first large-scale post-Six-Day War round of counterinsurgency warfare.

Upon its expulsion from Jordan in the autumn of 1970, in response to its abortive attempt to overthrow the Jordanian monarchy in a civil war, the PLO set up shop in Lebanon, leading to a second round of counterinsurgency warfare with Israel that would last until the organization's sweeping defeat in the 1982 Lebanon War. From the early 1970s through the early 1980s, the PLO resumed direct attacks on Israeli territory, as well as continued its international terror campaign. Israel responded to this new wave of hostilities with a robust air–land counterinsurgency campaign in Lebanon.[3] The IAF regularly flew interdiction strikes against PLO positions during these years, and it provided CAS to land forces, especially during the major IDF ground incursion into southern Lebanon in the spring of 1978 known as Operation Litani. Finally, the air force engaged in numerous reconnaissance sorties over Lebanon, transported special forces units behind PLO lines, brought supplies into the area, and medevaced casualties out of it.

This round of counterinsurgency warfare, however, did not yield the same positive results as the first round. While the IAF, not to mention the IDF's land forces, undoubtedly inflicted severe losses on the PLO's men and matériel, the organization's rocket attacks against and infiltration attempts into northern Israel were never brought to a complete halt. Indeed, it eventually took a full-scale war to smash the PLO's military infrastructure in Lebanon, as well as to oust the organization itself from the country (though remnants of some its factions remained behind to harass Israel over the following decades).

The demise of the PLO in Lebanon cleared the way there for the rise of a far more dangerous foe, Hizbullah. From the mid-1980s until Israel's withdrawal from southern Lebanon in the spring of 2000, the IAF was regularly in action against Hizbullah targets to the north of Israel's security zone.[4] In early 1992, helicopter gunships eliminated the organization's supreme commander in a "targeted attack," the first instance in which the air force carried out a strategic attack against a terrorist leader.[5] IAF aircraft flew large numbers of interdiction sorties against Hizbullah targets—rocket launchers, command posts, training bases, arms depots, and so on—throughout the 1980s and 1990s, especially during two serious escalations in the fighting, Operation Accountability in 1993 and Operation Grapes of Wrath in 1996. Air force helicopters frequently inserted special forces units into Hizbullah-controlled areas, and its unmanned aerial vehicles (UAVs) routinely conducted intelligence-gathering forays and battle damage assessments over the years.

The IAF's campaign certainly took a heavy toll on Hizbullah's men and matériel. And it helped to reinforce Israeli deterrence in the short term, particularly in the wake of Operations Accountability and Grapes of Wrath. Nevertheless, this round of counterinsurgency warfare did not produce the desired results, either. The IDF did not defeat Hizbullah, and Israel finally decided upon a withdrawal from southern Lebanon, despite the absence of a formal agreement to end cross-border hostilities.

Just a few months after Israel pulled out of southern Lebanon, the second intifada (or the so-called al-Aqsa Intifada) erupted in Judea, Samaria, and Gaza. During the first intifada in the late 1980s and early 1990s, the IAF had been largely inactive, its participation restricted to troop transport and casualty evacuation, because this round of hostilities took on the character of mob violence and isolated terrorist incidents rather than an organized insurgency. IDF infantry, together with the paramilitary border police, handled outbreaks of violence. The second intifada, to the contrary, quickly assumed the form of a classic insurgency; therefore, the IAF was called into the fray early on.

Repeated strategic attacks on the leadership of Palestinian terrorist organizations (Hamas, Islamic Jihad, Fatah, etc.) in the form of targeted attacks, often executed by AH-1 Cobra and AH-64 Apache helicopter gunships, clearly constitutes the IAF's greatest contribution to the eventual defeat of these groups in the second intifada. A significant number of senior- and middle-ranking terrorist commanders, including many of the most experienced homicide bomb dispatchers, who could not be replaced easily, were eliminated in these attacks.[6] In conjunction with other counterinsurgency measures—incursions by IDF land forces into areas of Judea, Samaria, and Gaza under the official control of the Palestinian Authority, construction of the security barrier along the Judean and Samarian border, and creation of a system of checkpoints and roadblocks—targeted attacks led to a precipitous decline in homicide bombings (not to mention other kinds of large-scale terrorist incidents), which had reached their height in 2002, until they had mostly died out by the time of the Israeli withdrawal from Gaza in the autumn of 2005.

The IAF also engaged in sporadic interdiction strikes, mostly executed by helicopter gunships, against terrorist command posts, training depots, and weapons manufacturing facilities in Judea, Samaria, and Gaza, as well as more frequent interdiction strikes against individual rocket launchers in northern and central Gaza during the second intifada. The air force flew a small number of CAS sorties on behalf of IDF land formations, again executed by helicopter gunships, during Operation Defensive Shield in 2002, an operation that spelled the beginning of the end for Palestinian terrorist organizations in Judea and Samaria, as well as during similar incursions carried out in Gaza prior to the Israeli withdrawal.[7] Collectively speaking, these interdiction and CAS sorties also made a contribution to the defeat of the Palestinian terrorist organizations.

Round-the-clock employment of UAVs for reconnaissance ensured that most of the IAF's strategic attack, interdiction, and CAS sorties were delivered with a high degree of accuracy and minimal collateral damage. In terms of its ancillary roles, in addition to flying innumerable reconnaissance sorties, the IAF also used its helicopters extensively for troop transport and casualty evacuation.

A Summary and Assessment of IAF Activity in Special Operations Warfare

Since the Six-Day War, the IAF has been a key player in at least four major special operations.[8] Three of these operations—two against Arab nuclear weapons facilities in Iraq and Syria, respectively, and one against PLO headquarters facilities in Tunisia—fall under the rubric of strategic attack. The other—an anti-terrorist hostage-rescue mission at Entebbe airport in Uganda—falls primarily under the rubric of transport.

To bolster its quest to become a hegemonic power in the Middle East, Iraq in the 1970s eagerly sought a nuclear weapons capability. To this end, it intended to use a French-built nuclear reactor outside of Baghdad to produce weapons-grade enriched uranium. Iraq's well-known antagonism to the existence of Israel, naturally, made the latter extremely wary of a nuclear weapons program in the hands of an archfoe. By the early 1980s, Israel's attempts to disrupt this program through a combination of covert operations and public diplomacy had seemingly reached the end of the line; therefore, Israel decided that a more direct and forceful response would be necessary to derail Iraq's program before it passed the point of no return.

Israel gave the IAF the task of impeding Iraq's nuclear weapons program. In June 1981, after a low-level, terrain-masking approach to the target, eight IAF F-16 Fighting Falcons dropped 16 2,000-pound general purpose "blockbuster" bombs on Iraq's nuclear reactor.[9] Fourteen of these bombs hit the target, reducing it to a heap of rubble within two minutes. In retrospect, the massive destruction caused by this strategic attack contributed mightily toward putting an end to Iraq's drive to attain a nuclear weapons capability. Iraq, to be sure, restarted its nuclear weapons program in the wake of the Israeli attack; however, the setback suffered as a consequence of the strike meant that it had not yet attained a nuclear weapons capability by the outbreak of the 1991 Gulf War, when the Allied Coalition once and for all stripped it of any semblance of a nuclear weapons program.

Syria, too, had its heart set on a nuclear weapons capability, especially in view of the fact that its stockpile of chemical arms had not given it "strategic parity" with Israel. Hence, with North Korean assistance, it constructed a weapons-manufacturing plant in the northeastern corner of the country, probably in an effort to hide it from Israel's prying eyes. This effort at concealment did not work, and once Israel confirmed the facility's true purpose—through a combination of satellite photography and ground reconnaissance—it decided to destroy the plant, in line with its long-standing policy of not permitting an archfoe to possess nuclear weapons.

Israel once again called upon the IAF to do away with this grave threat to its national security. In September 2007, Israeli fighter-bombers, probably F-15 Eagles, struck the facility, apparently with some type of precision-guided munition (PGM), most likely some sort of "bunker buster" bomb.[10] In the raid on the Iraqi nuclear reactor, the IAF achieved complete surprise by choosing the route to the target very carefully and by relying on the top-notch skills of its pilots. This time around, in contrast, the IAF seems to have had complete control of Syria's integrated air defense system (IADS) prior to the raid, possibly because a stealth UAV used the beam of one its air defense radars to implant code throughout the system that allowed the air force's cyber warfare practitioners to gain command.[11] Whatever the case, the IAF once more achieved complete surprise, and once more utterly demolished the target. This strategic attack, like its predecessor in Iraq, has proven to be an outstanding success, as Syria's nuclear program has not only been exposed before the international community, but it has also not been resumed in any serious way in the years since the raid.

The IAF's third strategic attack occurred in October 1985. Upon its expulsion from Lebanon in the summer of 1982, the PLO took up residence in Tunisia, from where it continued to orchestrate terrorist attacks against Israeli targets. In the wake of the murder of Israeli tourists in Cyprus, consequently, the IAF was instructed to attack a cluster of leadership targets in Tunis.[12] This most distant mission from Israeli shores in air force history, which required in-flight refueling, turned out to be a considerable tactical success, as the eight fighter-bombers on the raid hit all of their designated targets, causing substantial casualties among middle-ranking PLO commanders, while limiting the amount of collateral damage to the surrounding areas. Whether this strategic attack had more than a temporary effect in dissuading the PLO from engaging in further terrorist attacks, however, is not clear.

In addition to these strategic attacks, the IAF also played a pivotal role in perhaps the most spectacular hostage-rescue mission in history, the raid on Entebbe airport.[13] In the summer of 1976, terrorist operatives from one of the PLO's factions, together with terrorist operatives from an extreme left-wing German group, hijacked an Air France passenger jet with a large contingent of Israelis onboard. Following a brief stopover in North Africa, the jet flew on to Entebbe, where the hijackers enjoyed the protection of the Ugandan government and army. In exchange for the safe return of the hostages, the hijackers demanded the release of terrorist operatives imprisoned in Israel, Europe, and Africa. For the first few days of the incident, Israel thought seriously about giving in to the hijackers; however, once it became clear that the IDF could mount a raid on Entebbe, Israel chose to exercise its military option rather than to knuckle under to terrorism.

IAF C-130 Hercules transport aircraft flew special operations units to Entebbe under the cover of darkness, where they landed without drawing attention to themselves. An IDF special operations unit rapidly stormed the terminal building housing the hostages, killing the terrorist operatives, freeing the captives, and shepherding them to the waiting IAF transport aircraft, which would fly them back to Israel. Meanwhile, other special operations units eliminated several dozen Ugandan troops who tried to intervene in the fighting and blew up most of the Ugandan air force's fighter-bombers. The raid, which lasted about one hour, was accomplished for the loss of one Israeli soldier and three hostages. Beyond providing the transport aircraft, the IAF contributed an airborne hospital, which treated casualties transferred to it in Kenya, and an airborne command post that coordinated events on the ground during the raid.

It is difficult to measure with any degree of precision what impact the raid had on the international fight against terrorism. The raid probably did contribute to a long-term decline in the frequency of airline hijackings, and it certainly gave pause to anti-Israeli terrorist groups thinking about similar actions in the future. It also probably emboldened other states to take a firm stand against terrorism—for example, a West German counterterrorist unit successfully stormed a hijacked airliner in Mogadishu, Somalia, not long afterward.

Since the 1973 Yom Kippur War, however, the IAF has not only engaged in joint special operations with IDF land forces, as at Entebbe, but has also developed its own special operations units. In the aftermath of this war, the air force recognized that it needed an airborne rescue unit, as well as a reconnaissance/raiding unit. These requirements gave rise in the mid-1970s to Unit 669 and Shaldag, respectively. In later years, the IAF also created a number of additional small and highly specialized units under the aegis of its special forces command, but 669 and Shaldag have always been the jewels in its crown.

The IAF created Unit 669 primarily as a result of its experience with downed airmen in the Yom Kippur War.[14] Many of these airmen had parachuted behind Arab lines, and some of them had been injured before—or in the process of—abandoning their aircraft. With no dedicated airborne rescue unit at hand, the IAF had to create heliborne rescue teams on the spot. These improvised teams performed yeomanly work in recovering downed airmen, not to mention in airlifting wounded soldiers from the battlefields to rear-area hospitals; however, the sheer number of operations assigned to these teams convinced the IAF that it required a permanent rescue unit.

The members of 669—and they include a high proportion of doctors and medics—are highly trained special operations soldiers who undergo very rigorous training, as an abundance of the unit's rescue operations in the past have been performed close to—or behind—the front lines, often under heavy fire.[15] Many of 669's operations remain highly classified, because they involved the extraction of special operations units. Rumor has it that, during the air raid on the Osirak nuclear reactor, a 669 team infiltrated deep into the Iraqi desert via helicopter, where it concealed itself, just in case one (or more) of the aircraft involved in the attack were to be shot down. The prompt and effective medical care administered by this unit over the years has undoubtedly saved the lives of many airmen and soldiers.

Shaldag is the IAF's "deep penetration" reconnaissance/raiding unit.[16] Along with Sayeret Matkal, the IDF General Staff's special operations unit, and Shayetet-13, the Israel Navy's main special operations unit, it is considered one of the three finest "commando" units in the Israeli armed forces. Like members of Unit 669, troopers in Shaldag are highly trained in all aspects of special operations warfare. With the prominent exception of its participation in the airlifting of the Ethiopian Jewish community to Israel, Shaldag's operations are almost always shrouded in secrecy. During the 2006 Second Lebanon War, in one of its rare publicized operations, Shaldag, in cooperation with Sayeret Matkal, penetrated a heavily guarded Hizbullah compound deep in the Bekaa. As many as 20 terrorist operatives were killed and a number of others captured without loss to the Israeli raiding party.

Rumor has it that Shaldag also reconnoitered Syria's nuclear weapons facility before the IAF bombed it in the fall of 2007. Generally speaking, Shaldag operations involve either target designation for air-launched PGMs, scouting high-value targets behind the front lines, such as air bases, or raiding high-value targets behind the lines.

The IDF has recently decided to create a "depth corps," which would improve coordination among its deep penetration special operations units—including Unit 669, Shaldag, Sayeret Matkal, Shayetet-13, and, perhaps, others—in order to counter better threats to Israel emanating from far beyond its borders.[17] Whether these special operations units will be formally transferred to this new corps is unclear as of 2012, though the possibility certainly does exist that the IAF could lose control over 669 and Shaldag at some point in the future.

A Summary and Assessment of IAF Activity in Humanitarian Operations

A basic principle of Israeli foreign policy holds that the state has an obligation to rescue Jewish communities in distress around the world. In the early decades of statehood, endangered Jewish communities in Arab and East European states were brought to Israel via covert operations. The IAF did not play a role in these early rescue operations. To the extent that these operations involved flying refugees to Israel, they were brought in on chartered civilian aircraft.

This situation changed in the 1980s, when the air force played a significant role in bringing thousands of Ethiopian Jews to Israel via a covert airlift from Sudan. IAF C-130 Hercules aircraft landed clandestinely at makeshift airstrips in the Sudanese wilderness, where Jewish refugees were awaiting them after long treks from their home villages. Agents of Israel's foreign intelligence service, MOSSAD, as well as members of Shaldag, assisted in getting the refugees onboard the aircraft.[18] The airlift went on until it received exposure in Israel. This publicity prompted Sudan, which had tacitly cooperated with the operation, to shut it down. Over 8,000 Ethiopian Jews were brought to Israel during Operation Moses.[19]

By no means did this operation rescue the entire Ethiopian Jewish community. Indeed, the majority of its members remained behind to confront continued economic privation and political oppression. In 1991, the civil war then raging in Ethiopia presented an opportunity to bring the rest of the community to Israel. As one of its final acts, the dying Communist regime, with the tacit cooperation of rebel forces rapidly closing in on the capital of Addis Ababa, permitted Israel to mount an airlift. Employing its own sizable fleet of transport aircraft, as well as a number of commercial aircraft conscripted from El Al, the national airline, the IAF mounted perhaps the most spectacular human airlift in history, flying over 14,000 Ethiopian Jews to Israel in a day and a half during Operation Solomon.[20] One El Al Boeing 747 carried over 1,100 refugees to Israel, setting a record for the largest number of passengers ever carried on a single flight. Collectively, Operations Moses and Solomon transplanted most of the Ethiopian Jewish community to Israel.

On a much more modest scale, Unit 669 has also engaged in humanitarian operations since its creation in the aftermath of the 1973 Yom Kippur War.[21] During peacetime, its teams of highly trained special operations troops routinely rescue stranded hikers, climbers, and sailors from their predicaments across the Israeli landscape, and its doctors and medics transport the sick and injured to hospitals via specially equipped helicopters, treating them along the way. Pregnant Bedouin women in remote parts of the Negev Desert are said to prefer Unit 669 medics to hospital doctors and nurses when giving birth. Unit 669 looks upon these civilian rescue operations as supplementary training exercises that are useful to sharpening its wartime skills.

With the vast majority of the world's Jewish population residing in democratic and stable states, Israel's humanitarian operations have increasingly focused on helping non-Jews to recover from natural or manmade catastrophes. During its first few decades, Israel's humanitarian operations with respect to the wider world were largely restricted to training foreigners in such fields as medicine and agriculture, both at home and abroad, as well as to taking in small numbers of non-Jewish refugees from such places as Vietnam. With the growth of IDF disaster relief capabilities during the 1980s, Israel began to assist in the alleviation of overseas disasters, too. The IAF's role since the mid-1980s has been logistical in nature—to transport IDF medical and search and rescue (SAR) personnel, as well as basic necessities, to disaster areas. Over the past quarter century, the air force has participated in at least 15 humanitarian operations.

The IAF's first contribution to a relief mission occurred in 1985, when it transported medical and SAR teams—along with medical supplies, tents, and other necessities—to help Mexico recover from a destructive earthquake that hit its capital city.[22] In 1986, the IAF flew a medical team to Cameroon, where doctors, nurses, and medics treated thousands of persons suffering from respiratory ailments and chemical burns caused by a volcanic eruption. In 1988, the IAF airlifted a field hospital and SAR teams to Soviet Armenia to locate and care for earthquake survivors. The following year saw the IAF back in the Soviet Union, when it brought in a medical team to minister to survivors of a train wreck in the Ural Mountains.

This record grew extensively in the 1990s. In 1991 and 1992, the IAF transported disaster relief personnel and supplies to aid victims of a Georgian earthquake and the Yugoslavian civil war, respectively. The IAF participated in a larger-scale humanitarian operation in 1994, when it flew in a fully equipped, 120-bed field hospital to the Democratic Republic of Congo in order to assist victims of Rwanda's extremely bloody civil war. Israel was the first foreign power to respond energetically to this tragedy. Over the course of more than a month, the IDF field hospital treated thousands of patients, many of whom required major medical procedures.

In light of the number of humanitarian operations undertaken over the previous decade, the IDF, including the IAF, in cooperation with MASHAV, the Israeli government's civilian aid agency, set up a permanent aid unit in 1995 to streamline emergency response procedures. This new regime received its first big test in 1998, when al-Qaeda bombed the United States embassies in Kenya and Tanzania. The IAF brought in medical and SAR teams to Kenya. The SAR team pulled at least three survivors out of the rubble, while the medical team erected a clinic on site. Israeli medical personnel provided care to about 250 injured persons and performed at least eight major medical procedures before winding up their stay in Kenya.

In 1999, Israel responded to two massive disasters—one manmade, the other natural. During the civil war in Kosovo, the IAF transported a 100-bed field hospital to Macedonia to furnish medical care to refugees fleeing the war zone. In a two-week period, Israeli medical personnel, many of them veterans of the Rwanda and Kenya operations, treated more than 1,600 people, performed three major medical procedures, and delivered a dozen babies. Hard upon the heels of this operation, Israel responded to a massive earthquake in Turkey. The IAF flew in a SAR team and a fully equipped field hospital. The team, on the ground for a week, managed to pull 12 survivors out of demolished buildings at three different locations. The field hospital, on the ground for two weeks, treated over 1,200 patients and delivered 15 babies. The IAF also brought in materials for and experts in the rapid construction of dwellings. Indeed, Israel built an entire village consisting of 300 homes, a school, a medical clinic, and two playgrounds for 2,000 displaced Turks.[23]

Despite the second intifada, the 2006 Second Lebanon War, and the 2008–9 Gaza War (or Operation Cast Lead), Israel has continued to participate in humanitarian operations in the twenty-first century. In the aftermath of a powerful earthquake in western India, the IAF transported a 100-bed field hospital to the scene. Before they turned the hospital over to Indian medical personnel, IDF doctors, nurses, and medics treated over 1,200 patients and delivered a dozen babies. Israel offered the same level of assistance to Sri Lanka after the devastating tsunami in 2004; however, for logistical reasons related to its perceived inability to accommodate a large-scale foreign presence, the Sri Lankan government declined Israel's overture. Instead, an airlift brought in basic supplies, including food, water, medicine, blankets, tents, and generators. In 2006, the IAF flew a SAR team into Kenya to help dig survivors out of the rubble of a large building that had collapsed upon itself. The SAR team pulled two survivors out of the debris, and an accompanying IDF medical team cared for some of the injured.

In 2010, Israel was the first country to provide a comprehensive and coherent medical response to the earthquake in Haiti. While the aid efforts of other states floundered because of the tremendous destruction to Haiti's already primitive infrastructure, the IAF flew in a fully equipped field hospital. In its two weeks of operation before its equipment was turned over to Haitian locals, this field hospital treated over 1,100 patients, performed more than 300 surgeries, and delivered 16 babies. IDF SAR personnel also rescued many survivors from the rubble of buildings. Later in the year, the IAF delivered relief supplies to Colombia in the wake of a natural disaster in that state.

Israeli humanitarian efforts continued in 2011, with the IAF flying in a medical team to treat survivors of Japan's devastating earthquake and tsunami. This medical team administered care to several hundred patients over a two-week period, and then handed over its equipment to Japanese doctors and nurses. Toward the end of 2011, the IAF also brought relief personnel and supplies to Turkey in order to assist with recovery efforts from an earthquake in the eastern part of the state.

Conclusion

Israeli airpower, generally speaking, has been used rather effectively in counterinsurgency warfare and extremely effectively in special operations warfare. With respect to the former, the IAF played a substantial part in the defeat of two insurgencies since the Six-Day War, those that occurred in the late 1960s and the early 2000s. Airpower proved less effective against the Lebanon-based insurgencies of the 1970s through the 1990s; however, even in these cases, it inflicted serious damage on terrorist organizations and limited the amount of damage they were able to visit upon the IDF and Israel. With regard to the latter, the IAF countered two existential threats to Israel in the form of Arab nuclear weapons programs, and it dealt a couple of stinging blows to terrorism.

Partly as a consequence of the IAF's substantial capacity to airlift people and matériel, Israel has been able to punch well above its global weight in respect of participation in humanitarian operations. Though the era of large-scale humanitarian operations on behalf of endangered Jewish communities around the world may be over (at least for the foreseeable future), natural and manmade disasters are an ever-present part of the human condition. Given the speed and efficiency with which the IDF, including the IAF, has undertaken humanitarian operations in the past, Israel's services are sure to be among the first sought by stricken states and peoples in the future.

Top: The Mirage III served as the principal interceptor of the Israel Air Force (IAF) from the early 1960s to the mid-1970s. (Photographer: Moshe Pridan, © Israel Government Press Office)

Bottom: Mirage IIIs, including the one displayed here with eight air-to-air victories to its credit, shot down hundreds of Arab aircraft from the mid-1960s to the mid-1970s. (Photographer: Chanania Herman, © Israel Government Press Office)

Main photo: The F-4 Phantom spearheaded the IAF's strategic attacks during the 1973 Yom Kippur War. (Photographer: Moshe Milner, © Israel Government Press Office)

Above: The A-4 Skyhawk constituted the backbone of the IAF's close air support (CAS) effort during the Yom Kippur War. Today, it is relegated to the function of operational trainer, and it will soon be replaced by a more modern and suitable Italian jet aircraft. (Photographer: David Weinrich, © David Weinrich)

Opposite: Helicopters, like the UH-1 Iroquois shown here in the Yom Kippur War, have evacuated thousands of casualties from the battlefields of Israel's many wars. (Photographer: unknown, © Israel Government Press Office)

103

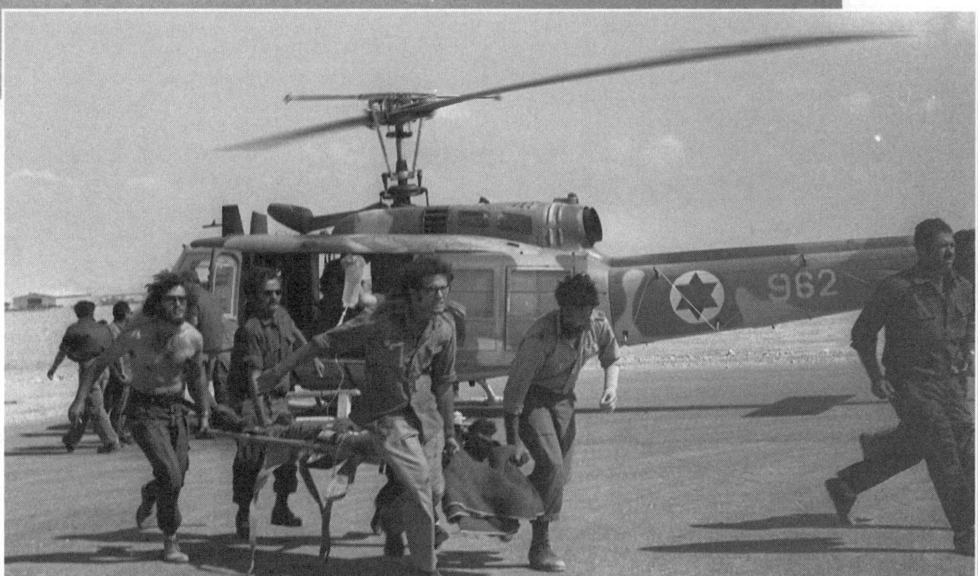
962

Above: The IAF has frequently used its helicopters to transport special operations forces behind Arab lines. This photo shows an SA-321 Super Frelon involved in the capture of the fortified Egyptian island of Shadwan during the 1969–70 War of Attrition. (Photographer: Moshe Milner, © Israel Government Press Office)

Below: Helicopter gunships, such as this AH-64 Apache, executed many CAS sorties in the 2006 Second Lebanon War and the 2008–9 Gaza War (or Operation Cast Lead). (Photographer: Tsvika Israel, © Israel Government Press Office)

Precision attacks by Apache and AH-1 Cobra helicopter gunships eliminated numerous terrorist operatives, including high-ranking leaders, during the second intifada (or the so-called al-Aqsa Intifada). (Photographer: Tsvika Israel, © Israel Government Press Office)

Main photo: The IAF employed midair refueling to extend the range of its F-15 Eagles so that they could destroy Palestine Liberation Organization (PLO) headquarters in Tunis, Tunisia, in 1985. (Photographer: Nati Harnik, © Israel Government Press Office)

Opposite: The raid on Entebbe airport, Uganda, in 1976 would not have been possible without the extraordinary capabilities of the C-130 Hercules. This aircraft has also been a prominent player in the IAF's contribution to Israel's humanitarian relief operations over the years. (Photographer: Moshe Milner, © Israel Government Press Office)

140

4X-FBQ
420

Main photo: The F-15I, along with the F-16I, is the IAF's principal strategic attack aircraft. (Photographer: David Weinrich, © David Weinrich)

Above: The F-15 Eagle remains the IAF's most potent aircraft in the realm of air-to-air combat. Even the acquisition of the F-35 Lightning II is unlikely to displace it in this regard. (Photographer: Moshe Milner, © Israel Government Press Office)

Opposite: The F-16 Fighting Falcon currently equips most of the IAF's fighter-bomber squadrons. (Photographer: David Weinrich, © David Weinrich)

238

129

The F16I—the latest version of this aircraft to enter IAF service—has a longer range and much improved ordnance and avionics capabilities in comparison to earlier models. (Photographer: David Weinrich, © David Weinrich)

The IAF's drone fleet provided copious, real-time intelligence information during the second intifada, the Second Lebanon War, and the Gaza War.
(Photographer: Yaacov Saar, © Israel Government Press Office)

The Hawk missile system first entered IAF service in the mid-1960s, and it has downed dozens of Arab aircraft since it became operational. It is slated to be replaced by the David's Sling interceptor system in the coming years. (Photographer: Moshe Milner, © Israel Government Press Office)

The Patriot missile proved unable to intercept Iraqi ballistic missiles during the 1991 Gulf War, but it remains a potent component of the IAF's integrated air defense system (IADS). (Photographer: Moshe Milner, © Israel Government Press Office)

Above: The Soviet SA-3 caused a lot of problems for the IAF when it first appeared on the battlefield during the 1969–70 War of Attrition. (Photographer: Gad Binter © Israel Government Press Office)

Left: Israel's opponents have obtained large numbers of ballistic missiles, some of which are armed with weapons of mass destruction. The Arrow anti-ballistic missile interceptor system is meant to defend the state against this threat. (Photographer: Israel Aerospace Industries, © Israel Government Press Office)

The Ofeq class of reconnaissance satellites, one of which is being launched into space atop this Shavit rocket, has given the IAF the capability to gather intelligence on Israel's most distant foes. (Photographer: Israel Aerospace Industries, © Israel Government Press Office)

4

An Airborne Revolution in Military Affairs: Unmanned Aerial Vehicles in Israel Air Force Service

Unmanned aerial vehicles (UAVs), often referred to colloquially as drones, can claim a lineage that dates back to the dawn of air warfare. Though quite rare in comparison to the enormous numbers of manned aircraft involved in the first and second world wars, UAVs participated in both conflicts, especially the latter, mainly as attack vehicles armed with high-explosive warheads. Not until the Vietnam War, however, did drones really find a defined niche on the battlefield, when the United States Air Force conducted thousands of reconnaissance sorties over hostile territory with UAVs. With the possible exception of the United States, the State of Israel is the polity most closely identified with UAV operations in the post-Second World War period. It has actually employed drones in a variety of roles since the early 1970s, but it initially gained worldwide attention for its operations during the 1982 Lebanon War, in which its UAVs played a substantial part in the destruction of the Syrian integrated air defense system (IADS) erected in Lebanon. Recent asymmetrical wars—the 2006 Second Lebanon War and the 2008–9 Gaza War (or Operation Cast Lead)—sparked renewed global interest in Israeli drone operations.

Nevertheless, outside of the international defense community—professional soldiers, military analysts and journalists, arms designers, and so on—familiarity with Israel's UAV operations, past and present, is not widespread. A brief review of its experience with drones, as well as a few thoughts about the future of its UAV force, then, seems entirely in order, particularly as the employment of these vehicles is set to expand dramatically in the years ahead, if the fighting in places as diverse as Afghanistan, Pakistan, Iraq, Lebanon, and Gaza is any indication of what is just over the horizon.

UAVs and the Arab–Israeli Conflict

The Israel Air Force (IAF) first employed UAVs on a large scale in the 1973 Yom Kippur War.[1] During the opening days of the conflict, the air force suffered heavy losses at the hands of the Egyptian and Syrian IADSs. Instead of focusing its efforts on the destruction of these systems at the outset of hostilities, as envisaged in its prewar operational plans, the IAF was called upon to stem the advance of Egyptian and, especially, Syrian land forces in the Sinai and on the Golan, respectively, because the Israel Defense Forces (IDF), taken by surprise by the Arab assault, had not deployed to the fronts its reserve units, which constituted the bulk of its fighting power. The IAF, therefore, flew hundreds of interdiction and close air support (CAS) sorties against the Egyptian and Syrian armies, regardless of the cost exacted by their IADSs.

Once the IDF's reserve units reached the battlefields, blunted the Egyptian and Syrian offensives, and stabilized the fronts, the IAF sought to cut its losses to anti-aircraft fire. In its quest to do so, it began to employ its American-supplied Firebee and Chukar drones—Israel had yet to deploy any UAVs of indigenous design and manufacture—as decoys to draw this fire away from its aircraft, especially on the Sinai front. The fact that the IAF's losses to anti-aircraft fire dropped dramatically after the first few days of hostilities suggests that the drones had a positive impact on the air war. Whether the IAF also employed its UAVs to gather photographic intelligence and to attack surface-to-air missile (SAM) and anti-aircraft artillery (AAA) batteries is not known. In any case, its Yom Kippur War experience appears to have convinced the IAF that drones could be effective tools on the battlefield.

During the late 1970s, Israel fielded its first generation of home-grown UAVs, the Scout and Mastiff. Though small and unsophisticated by the standards of later generations of drones—these compact, twin-tailed, propeller-driven vehicles carried very limited payloads of rather simple electronic systems, mainly video cameras and, perhaps, infrared detection equipment—they nevertheless proved quite effective in service. Prior to the Lebanon War, these vehicles, in tandem with IAF reconnaissance aircraft, routinely monitored the Syrian IADS in Lebanon.[2] While expendable decoy drones drew anti-aircraft fire—a few of them were even shot down—other drones and reconnaissance aircraft gathered valuable information on the locations and electronic signatures of SAM batteries, which the IAF then integrated into its operational plan for destroying the Syrian IADS in a potential future confrontation. With the commencement of hostilities in summer 1982, UAVs played a prominent part in the IAF's spectacular elimination of the Syrian IADS in the Bekaa, which saw about 20 SAM batteries knocked out on the first day of Operation Mole Cricket 19, the code name given to the plan to demolish that air defense network. Subsequent air strikes on following days wiped out additional SAM batteries. The IAF also destroyed considerable numbers of AAA batteries.

The IAF's drones filled several roles during the battle. First, decoy UAVs, especially the locally developed Samson, lured the Syrians into activating their radar systems, which then fell prey to air-delivered precision-guided weapons (PGMs), such as the American Standard anti-radiation missile (ARM), as well as ground-launched PGMs, such as the Israeli Keres ARM. Other air-delivered PGMs, like the Israeli Tadmit television-guided missile and the American GBU-15 television-guided glide bomb, added to this maelstrom.[3] With the radar systems out of commission, aircraft and artillery smashed the SAM launch positions at will with a mixture of general-purpose bombs and shells, as well as cluster munitions. Second, one SAM battery may actually have been taken out by a UAV fitted with a warhead, perhaps a precursor to—or prototype of—the later Israeli Harpy attack drone.[4] Third, the IAF undoubtedly employed UAVs for real-time surveillance and target acquisition during the battle, as well as for post-battle damage assessment.

During the Lebanon War, UAVs also engaged in other missions on behalf of the Israeli war effort. They provided constant, real-time surveillance of Syrian air force bases, alerting IAF air battle controllers to the take offs of Syrian aircraft. This information helped the controllers to vector IAF F-15 Eagles and F-16 Fighting Falcons to optimal intercept coordinates, contributing to the lopsided score in a series of air battles, in which Israeli pilots shot down 80–100 Syrian aircraft without incurring a single loss. One UAV even scored a "no weapons kill" of its own through wild maneuvering, when a Syrian aircraft attempting to shoot it down collided with the ground after the pilot lost control.[5]

Finally, UAVs assisted the IDF's land campaign, too. Drones furnished real-time intelligence on the location and movement of Syrian and Palestine Liberation Organization (PLO) units. Such data clearly assisted IDF commanders in planning and executing impressive tactical engagements, such as the large-scale defeat inflicted on Syrian armor by Israeli tanks and infantry around Lake Karoun. The employment of drones as part of the IDF's land campaign, in short, opened up a whole new avenue in air–land battlefield cooperation.

Buoyed by the accomplishments of its UAV force in the Lebanon War, Israel has continued to develop more sophisticated and specialized drones over the following decades. During the 1990s, the IAF deployed both the Searcher 1 and 2, essentially significantly bigger and more capable versions of the earlier Scout and Mastiff, fitted with broader and more advanced arrays of electronic systems, as well as the Harpy attack drone intended primarily to destroy air defense radar systems. In the same decade, Israel tinkered with the idea of developing a long-range, missile-launching drone to shoot down ballistic missiles in their "boost phase," but apparently abandoned the effort on cost grounds.[6]

Over the past decade, the IAF has fielded the Hermes 450, Hermes 900, Heron, Heron TP, and, possibly, other (classified) UAV models. The Heron TP is an especially large vehicle—it has the wingspan of a Boeing 737 aircraft—that can carry a state-of-the-art suite of sensors, including electronic warfare systems. Some of these drones can apparently be armed with small missiles, such as the American Hellfire or the Israeli Spike, for attack missions.[7] And the Heron TP is supposedly able to carry a one-ton bomb. Israel's aerospace industry has also developed several models of hand-launched "micro-UAVs," a number of which have been field-tested by IDF infantrymen for short-range intelligence-gathering missions.

Naturally, the accumulated knowledge gained by Israel in UAV design and deployment has not gone unnoticed by other countries. Israeli drone technology has been exported around the globe over the past three decades. Countries like the United States, India, Turkey, Great Britain, and Germany either have bought UAVs directly from Israel or they have manufactured them at home under license. Furthermore, IAF drone operators routinely share their experiences with their foreign counterparts.[8] American drone operations in Iraq and Afghanistan in particular have benefited heavily from Israeli input.

In the 1980s and 1990s, the main operational theater of the IAF's UAV force remained southern Lebanon.[9] The air force's drones played an especially active part in Israel's two large-scale anti-Hizbullah escalations of the fighting, Operation Accountability in 1993 and Operation Grapes of Wrath in 1996. Even before these escalations, the IAF employed UAVs to locate Hizbullah training camps, arms depots, command posts, and rocket launchers. During the fighting, drones provided real-time data on various targets, including moving vehicles in the process of transporting terrorist operatives from one location to another, as well as rocket launch sites, to air and artillery units, which then engaged these objectives with precision fire. It is unclear whether

UAVs also "lit up" targets with laser designators for air-delivered PGMs to home in on. It is certain, on the other hand, that the round-the-clock, real-time intelligence furnished by drones proved effective in many instances in knocking out Hizbullah targets, even if the operations themselves ended inconclusively.

The participation of the IAF's UAV force in IDF operations has grown ever larger in the latest rounds of Arab–Israeli hostilities, namely the second intifada (or the so-called al-Aqsa Intifada), the Second Lebanon War, and Operation Cast Lead.[10] During the intifada, drones, most prominently, relayed real-time intelligence on terrorist positions and movements in Judea, Samaria, and Gaza to air and land units around the clock. Though no information has been made public on specific UAV missions, it is likely that drones played an integral part in many "targeted attacks" by AH-1 Cobra and AH-64 Apache helicopter gunships on terrorist operatives and Qassam/Grad rocket launch sites. These attacks, collectively speaking, killed considerable numbers of high-ranking terrorists and disabled many Qassam/Grad batteries. Armed drones may even have executed some of these targeted attacks.

Perhaps the most extensive deployment of drones, however, occurred during Israel's two recent asymmetrical wars—the Second Lebanon War and Operation Cast Lead. Early on in the Second Lebanon War, IAF aircraft essentially destroyed Hizbullah's long-range rocket force in 30 minutes of intensive air strikes.[11] UAVs, quite likely, not only helped to pinpoint the launch vehicles prior to this air assault, but also took part in target acquisition during the strikes and in battle damage assessments after them. Real-time surveillance of medium- and short-range rocket launch sites by drones throughout the fighting also drastically shortened the "sensor-to-shooter" loop by the end of the war, with IAF aircraft and helicopter gunships able to destroy launchers within a mere two minutes of launch detections by UAVs. Armed drones may have carried out some attacks on Hizbullah targets, while other UAVs may have "painted" these objectives with laser designators for air-delivered PGMs.

Drone deployment in Operation Cast Lead more or less followed the pattern established in the Second Lebanon War. The IAF's UAV force primarily provided real-time intelligence to air and land units for them to act upon in engaging Hamas targets, whether command posts, transport vehicles, rocket launchers, arms storage depots, or even individual terrorist operatives. Armed drones may once more have executed some strikes with Hellfire or Spike missiles, while other UAVs may have actively assisted aircraft or helicopter gunships by lighting up targets for PGMs. And, in a new wrinkle intended to minimize collateral damage, drones armed with small, nonlethal "roof-knocker" missiles—that is, missiles armed with "flash-bang" warheads—were employed to disperse without injury human shields cynically employed by Hamas in order to deter attacks against what were unambiguously military targets.[12]

Interestingly, around the time of Operation Cast Lead, unconfirmed news reports surfaced to the effect that IAF drones had engaged in sorties very far from Israel's borders. One story had them conducting electronic warfare missions over Iran in order to interfere with ballistic missile tests by jamming telemetry systems. Another story had them flying surveillance and battle-damage-assessment missions over Sudan in support of an air strike there that destroyed a Hamas-bound Iranian arms convoy. The Heron and Heron TP drones are capable of staying in the air for at least 24–36 hours (most likely much longer), which means that they certainly have the range to engage in such missions.

UAVs and the IAF's Future Force Structure

In terms of the technological sophistication of its UAV force, Israel is unquestionably well ahead of the pack. Only the United States is in the same league. Likewise, in terms of the scope and effectiveness of the tactical uses to which it has put its UAV force, Israel is far ahead of other countries. Again, only the United States is a peer in this regard. Nevertheless, even though the IAF considers drones to be an extremely valuable and cheap "force multiplier," it has not divulged any concrete information about whether it intends to develop its UAV force into an instrument that will equal, or even surpass in certain respects, its manned aircraft fleet in the coming decades.[13]

The IAF, according to foreign reports, presently fields three drone squadrons in its order of battle.[14] These same sources credit it with possession of 15 fighter-bomber squadrons, four helicopter gunship squadrons, as well as a whole range of transport, training, intelligence-gathering, and electronic warfare squadrons. Moreover, the air force intends to add to its arsenal in the next decade from one to three fighter-bomber squadrons of the new F-35 Lightning II fifth-generation stealth aircraft (which would probably replace its oldest F-16s). The bottom line here is that, for the near future anyway, the air force order of battle appears as if it will be weighted heavily in favor of manned aircraft.

Such an emphasis would seem to make sense for now. Drones, after all, will not possess the capabilities to fulfill many of the roles of manned aircraft for years to come, and may never be able to fill some of them. UAVs (with the potential exception of the very largest among them) cannot presently carry the types of—not to mention the quantities of—bombs and missiles necessary to destroy large, heavily fortified targets, such as Iran's nuclear installations. Nor can they defend airspace against hostile aircraft and drones. Nor can they move soldiers and supplies to, from, and around the battlefield or deep into an opponent's hinterland. Nor can they provide as thorough intelligence-gathering and electronic warfare coverage as dedicated manned aircraft in some situations. The list of missions that drones either cannot yet execute at all or can only carry out less effectively than manned aircraft, of course, could be extended well beyond this handful of examples.

Still, the list of missions for which UAVs are fit has grown substantially over the past few decades, a reality to which the IAF is certainly sensitive.[15] Furthermore, not only are drones much cheaper to build, equip, and fly than manned aircraft, but they also do not risk the lives and limbs of the men and women who operate them. And, unlike manned aircraft, they can remain over a target area for long periods of time in order to gather intelligence, to disrupt communications, or even to launch an attack. They can fly lower to the ground than manned aircraft, too, making them better at gathering certain types of intelligence data. For these reasons alone, it would seem to make a great deal of sense for Israel not only to expand considerably its drone force in size and capabilities, but also to formulate a general operational doctrine for their employment alongside its manned aircraft fleet. Whether in a full-scale conventional war, an asymmetrical war, or a counterinsurgency campaign, a massed UAV force able to swarm the battlefield with whole squadrons of drones dedicated to specific roles—such as attack, reconnaissance, and electronic warfare—would clearly be of tremendous assistance to the IAF in achieving Israel's military objectives.

The IAF, naturally enough for an air force of a state perpetually in conflict with its neighbors, is a very security-minded organization, so its long-range work plan with respect to building up the numbers and capabilities of its drone squadrons is a carefully guarded secret. Similarly, though it is likely that the air force has given sustained, in-depth thought to the formulation of a general operational doctrine for their employment, its thinking on this issue as well is a closely held secret.

For clues, albeit vague and sporadic ones, as to the direction in which the IAF may be headed in the realm of UAV warfare, however, one can turn to Israel's premier aerospace think tank, the Fisher

Institute for Air and Space Strategic Studies, which is the research arm of the Israel Air Force Center (IAFC), a nongovernmental organization with close ties to the air force. The institute has a study center devoted to thinking about how drones can be employed on current and future battlefields.[16] Much of the Fisher Institute's research on UAVs is also secret, but it has published some literature on drone warfare and has hosted at least one conference partially open to the public on the topic. At this gathering, active and retired senior IAF officers indicated that, while the air force does not yet have quite enough confidence to entrust drones with its most vital missions, it is definitely committed to the process of making them an ever larger part of future operations, including CAS, interdiction, and strategic attack sorties.[17]

UAVs and Israel's Strategic Landscape

A large and robust UAV force has the decided potential to alter Israel's strategic landscape in the future, with implications for both nuclear and conventional deterrence. In respect to the former, drones could buttress deterrence in two distinct ways. First, a sizable fleet of long-range, highly advanced reconnaissance drones could supplement its growing surveillance satellite capabilities. The ability of UAVs to hover over a target area for long periods means that they could effectively serve as "gap fillers" for space-based platforms that cannot remain "on station" for the same duration of time.[18] This capacity would be particularly valuable during a crisis situation. Drones, for example, could linger over Iranian ballistic missile launch sites, continually monitoring the goings on there. If linked to active (e.g., the Arrow anti-ballistic missile interceptor) and passive (e.g., warning sirens) defenses, UAVs could enhance Israel's ability to counter a nuclear (or biological or chemical) weapons strike with ballistic missiles. Aware of this capability, a potential attacker, such as Iran or Syria, would have to add this consideration to its decision-making calculus before deciding upon a strike.

Second, drones could enhance Israel's nuclear deterrence by reinforcing its "second strike" capability.[19] Though Israel, according to media reports, already possesses a potent retaliatory capability—one based upon a triad of submarine-based, nuclear-armed cruise missiles; land-based, nuclear-tipped ballistic missiles; and air-delivered nuclear bombs—UAVs could nevertheless bolster this asset. Drones configured either to deliver nuclear weapons (e.g., via air-launched cruise missiles) or to serve as the actual attack vehicles (e.g., as super-smart,

That is what a cruise missile is!! Fuck!

super-destructive descendants of the German V-1) could strengthen the aerial component of the triad. UAVs, because they require much less ground infrastructure than manned aircraft, and can therefore fly from comparatively small and remote sites that would be hard to locate, let alone hit, would be highly likely to survive a first strike. Once again, a potential attacker would have to take this consideration into account before deciding upon a strike.

In respect to conventional deterrence, drones could augment Israel's overall qualitative edge over its opponents on the battlefield in the decades ahead. Many experts maintain that fifth-generation aircraft, such as the F-35, essentially represent the end of the line for manned fighter-bomber development. Though UAVs may never entirely displace manned aircraft over the battlefield, they are in many ways the wave of the future with respect to aerial warfare. This trend augurs rather well for Israel, as it is presently far ahead of its opponents in the realm of drone warfare. Moreover, this technological and doctrinal gap, in all likelihood, will only grow wider in the coming years, as Israel is in a position to expand its capabilities in this area at a much faster pace than its opponents.

Finally, over time, UAVs could modify somewhat the dynamics of the American–Israeli patron–client relationship. The national security threats faced by Israel are quite long-standing, multifaceted, and serious. This dangerous political environment, coupled with the state's limited economic capacity, means that it will remain dependent on American security assistance indefinitely; however, the build up of a large and powerful drone force could lessen the magnitude of that dependence. Israel, after all, is most dependent on the United States in regard to the supply of advanced fighter-bombers. To the extent that these aircraft are replaced in the future by indigenous UAVs, there will be a concomitant reduction in its reliance on American assistance. Less dependence on security assistance from the United States, in turn, could lead to greater freedom of action for Israel in a crisis situation where its national interests and those of its benefactor do not necessarily coincide.

Still needs our money.

Conclusion

A recent IAF commander has summarized the air force's experience and attitude toward drones in the following way:

> In terms of flying hours, the UAV layout is already in second

> place in the IAF, after the fighter planes. All of the [drones] create many "fringe benefits" and open a window to new worlds in the concept of air operations. The UAVs do not [currently] replace manned aircraft, but generally supplement the forces operating in the air. They also possess unique capabilities, especially the length of time they can remain aloft.
>
> We're in an age of very interesting UAVs, which come in an assortment of sizes and types[,] and are able to carry heavy loads. In this area[,] Israel is a superpower, not only in UAV development but also in the invaluable experience it has gained in operating and exploiting these vehicles to their fullest. In other words, this is a self-nourishing cycle, and I'm certain that we'll continue to develop in the world of UAVs just as we have up to now.[20]

Drones, in other words, will become an even bigger and more integral part of the IAF's arsenal and operations in the years and decades to come.

5

Ground-based Air Defense and Space-based Reconnaissance: Other War-related Responsibilities of the Israel Air Force

The Israel Air Force (IAF) has contributed to the defense of the State of Israel primarily by employing its aircraft and, more recently, unmanned aerial vehicles (UAVs) over the battlefield. This basic fact, however, should not obscure another salient fact—that the air force has also contributed to Israeli security in two other capacities. First, it has been in command of all of the state's ground-based air defense assets, both anti-aircraft artillery (AAA) and surface-to-air missile (SAM) batteries, since late 1970. And, second, it has been in charge of the state's reconnaissance satellites, the first of which entered into orbit in the late 1980s.

Ground-based Air Defense

Israel acquired its first AAA batteries during the course of the 1947–49 War of Independence.[1] The Israel Defense Forces (IDF) employed its handful of AAA batteries mainly to defend point targets of particular importance to the new state. Though they managed to knock down at least one Egyptian Air Force Spitfire as it attacked an IAF air base, the contribution of these batteries to the overall Israeli war effort was quite modest at best.

From the end of the War of Independence until the 1967 Six-Day War, the IDF's AAA batteries had few opportunities to engage the enemy. In the 1956 Sinai Campaign, they did not register any victories over Egyptian aircraft. Rather, IAF interceptors engaged many of the

Egyptian aircraft that attempted to fly close air support (CAS) or interdiction sorties against IDF land forces, shooting down a number of them in air-to-air encounters.

Nevertheless, the IDF continued to pour resources into the development of an integrated air defense system (IADS) throughout the 1950s and '60s. By the outbreak of the 1967 Six-Day War, the IDF had approximately 550 AAA batteries, 200 of them radar-guided, along with the associated command and control (C^2) infrastructure.[2] Interestingly, while the IDF's land forces retained authority over the AAA batteries, the IAF fought for—and won—control over Israel's first Hawk SAM batteries, which arrived in 1965.[3] During the Six-Day War, this bifurcated and rudimentary IADS managed to knock down three Arab aircraft.[4] The Egyptian, Jordanian, and Syrian air forces were largely destroyed on the ground during the first day of the war—and many of their few surviving aircraft were subsequently engaged and destroyed by IAF interceptors before they could attack IDF land forces—so the minimal score registered by Israeli ground-based anti-aircraft defenses is not too surprising.

The Israeli IADS improved its score to 15 Egyptian aircraft during the 1969–70 War of Attrition, at least some of them shot down by Hawk SAM batteries.[5] To improve further the performance of its IADS, the IDF decided in late 1970 to consolidate all of Israel's anti-aircraft defenses under the aegis of the IAF. By the outbreak of the 1973 Yom Kippur War, the IAF had under its control approximately 1,000 AAA batteries, 200 of them radar-guided, 12 Hawk SAM batteries, as well as an enlarged and upgraded C^2 infrastructure.[6]

In the Yom Kippur War, the Israeli IADS shot down 50–100 Arab aircraft, as many as 22 of them falling to Hawk SAM batteries.[7] The much higher score—in contrast to previous Arab–Israeli wars—is attributable to the fact that, in this war, the Arab states had the initiative at the outset of the fighting and committed their air forces over the battlefield in a considerably more aggressive fashion than in the past. Whatever the precise figure of Arab aircraft destroyed by anti-aircraft fire, the IDF decided to expand its IADS in the aftermath of the war. To this end, the IAF added to and modernized its inventory of ground-based air defense systems by acquiring Improved Hawk SAM batteries, mobile Vulcan AAA batteries, mobile Chaparral SAM batteries, as well as man-portable Redeye and, later, Stinger SAMs.

The next test of the Israeli IADS took place during the 1982 Lebanon War.[8] In this conflict, the IDF once again had the initiative. The IAF swiftly smashed the Syrian IADS in the Bekaa, and prowling Israeli interceptors shot down scores of Syrian aircraft sent up to protect it; therefore, IDF land forces rarely came under air attack during the fighting. Nevertheless, a Vulcan AAA battery shot down at least one ground attack aircraft, an Improved Hawk SAM battery helped to shoot down a high-flying Mig-25 (the first occasion on which this reconnaissance aircraft had been hit by ground fire), and a Redeye SAM damaged at least one aircraft. Parenthetically, Vulcan batteries also proved to be very effective in providing fire support to advancing land forces, especially in urban areas.

From the early 1990s forward, the Israeli IADS has focused more on ballistic missiles and rockets than on aircraft. In recent decades, after all, the threat posed by conventional warfare has receded somewhat, while the threat posed by terrorism and weapons of mass destruction (i.e., ballistic missiles armed with nuclear, chemical, or biological warheads) has increased considerably. The first manifestation of this trend could be seen in the 1991 Gulf War, when Iraq fired approximately 40 ballistic missiles at Israel throughout the course of the conflict. At the time, Israel had no active defense system able to counter these missiles. Hence, it had to rely on hastily deployed American and Dutch Patriot SAM batteries that ostensibly had the capability to destroy ballistic missiles. Jointly operated by American, Dutch, and Israeli crews, Patriot SAM batteries initially appeared to be quite effective at destroying the missiles; however, a thorough postwar analysis revealed that they had not, in fact, intercepted a single incoming missile.[9]

Israel had actually begun to develop an anti-ballistic missile interceptor system prior to the Gulf War, but it was nowhere near operational status in early 1991. The nightmarish experience of coming under ballistic missile fire—which caused considerable property damage, but few casualties (thanks to an efficient passive defense effort and some luck)—spurred Israel to speed up development of its Arrow anti-ballistic missile interceptor system (and its associated Green Pine radar and Golden Citron control unit). The Arrow system achieved initial operational capability in late 2000, with the first battery deployed at an air force base in central Israel.[10] Presently, the IAF has two—possibly three—batteries operational, and the system will be upgraded in the near future with a more capable interceptor, a more capable radar unit, and a more advanced command, control, communications, and intelligence (C^3I) unit.[11] In tests over the years, the Arrow system has proven itself to be quite efficient at destroying simulated ballistic missiles, and the enhanced system will feature the capabilities to engage effectively salvos of missiles and to engage first the most threatening incoming targets. An IAF officer in charge of administering the Arrow project observed the following about the upgraded interceptor:

> An additional specialty of the missile [is] its high-level targeting capabilit[y], which enable[s] the launching of [a] missile to [a] standby point in outer space, and once the target is clearly identified, the missile can intercept it. . .[12]

The Arrow system represents the top tier of the multilayered Israeli IADS. The middle tier of this IADS is currently filled by Patriot and Improved Hawk SAM batteries, as well as AAA batteries, many of which defend strategic facilities throughout Israel, such as air force bases and the nuclear reactor at Dimona.[13] Currently, these SAM and AAA batteries are limited to an anti-aircraft role; however, in 2013–14, the IAF will field the David's Sling—sometimes also called the Magic Wand—interceptor system that will not only be able to shoot down aircraft, but also medium-range rockets and cruise missiles. An IAF officer in charge of administering the David's Sling project remarked that:

> [The system] is based on highly advanced, cutting-edge interception technology. . . . It is a highly accurate system that can operate in any weather conditions[.] [I]t both analyzes and intercepts the threat. . . . [Its] area of protection is very wide. It is not a battery-based system, but rather [is] located in a central area, with a few additional sites, which allow it to cover the entire country. . .[14]

The David's Sling system is slated to replace the IAF's Improved Hawk SAM batteries.

Finally, the lower tier of the Israeli IADS is now filled by the Iron Dome system, which is intended to destroy short-range rockets, such as the Grad (or "Katyusha"), and long-range artillery shells. Terrorist organizations like the Palestine Liberation Organization (PLO) and Hizbullah have bombarded Israel with rockets for decades; however, it took the thousands of rockets that hit the northern part of Israel in the 2006 Second Lebanon War, as well as the thousands of rockets that hit the southern part of the state from Gaza beginning in 2001, to convince Israeli defense planners to authorize development of an interceptor system optimized to destroy short-range rockets. Indeed, just as it had initially opposed the development of the Arrow system, fearing (incorrectly in hindsight) that its cost would detract from the armed forces' ability to deploy offensive weapons systems, the IDF, including the IAF, opposed at first the development of such a purely defensive system, but eventually relented under governmental and public pressure.[15]

After investigating and rejecting systems based on high-energy lasers and rapid-fire guns, Israel developed the Iron Dome system in a mere four years.[16] The IAF deployed at least three batteries in southern Israel in 2011, and it plans to field many more batteries to defend both the northern and southern parts of the state in the not-too-distant future. Of the three interceptor systems, it is the only one that has been tested in combat as of 2012. In a few rounds of fighting in 2011 and 2012, the system, which only intercepts rockets heading toward populated areas, shot down 75–93 percent of engaged targets, despite the inevitable teething troubles.[17] Even if the low figure is accurate, the Iron Dome system must thus far be reckoned a success, as short-range rockets are notoriously difficult to knock down in flight. Indeed, the system passed its first large-scale test with flying colors in late 2012 by shooting down hundreds of Hamas (and allied terrorist organization) rockets aimed at Israel's southern and central population centers, maintaining a "kill" rate of approximately 90 percent throughout the fighting across the Gaza frontier.

The IAF has already added an upgraded Iron Dome system to its inventory. This improved system is equipped with a better interceptor missile in larger numbers, as well as a more capable radar, giving it wider area coverage.[18]

Space-based Reconnaissance

In the Yom Kippur War, the United States apparently refused to supply Israel with satellite imagery of the battlefields, even though the Soviet Union shared such intelligence with its Arab allies.[19] This denial of access to vital information served as the immediate cause of the Israeli decision to embark upon a satellite development program. Israeli defense planners not only intended to achieve as much independence as possible in the sphere of intelligence collection, but they also concluded that Israel would need "eyes in the sky" in the coming decades in order to monitor emerging foes like Iran that were not located adjacent to the state's borders.

To these ends, Israel began to develop its first intelligence satellite in the early 1980s—once more over the objections of the IDF, including the IAF, which feared the budgetary consequences (again, incorrectly in hindsight). In contrast to the large and expensive American and Russian satellites that rest in high Earth orbits, Israel decided to build smaller and cheaper satellites that could be deployed on short notice in low Earth orbits equipped with optical and radar technology tailored to its specific requirements.[20] In mid-1988, Ofeq 1 was successfully launched into orbit atop the Shavit booster rocket, making Israel one of only a handful of states in the world with the capability to put satellites into space.[21] Ofeq 2 followed in early 1990. These two satellites did not produce intelligence photos, but rather served only as test vehicles.

Ofeq 3, placed into orbit in 1995, transmitted the first intelligence photos to the IAF's satellite control center, but ran out of fuel several years ago. Currently, Ofeq 5, 7, 8, and 9 are delivering photos to the IAF. Ofeq 5 will be replaced by Ofeq 10 in the coming years. The payload of Ofeq 8, also known as TecSAR, carries a synthetic aperture radar (SAR) system, which provides an all-weather, day/night intelligence capability. The precise optical and radar capabilities of these satellites, of course, are closely guarded secrets; however, it is believed that the range and quality of the images has improved with each successive deployment. In contrast to American and Russian satellites, which make only one or two passes per day per satellite over the Middle East, each Ofeq satellite makes up to six passes per day over the area, giving Israel better real-time satellite coverage of the region than either of these states.

Today, Israel is far along the road toward developing microsatellites weighing 100 kilograms or less that can be launched into space from airborne platforms like fighter-bomber aircraft.[22] These very small and cheap satellites would be deployed in low Earth orbits "on demand" in order to reconnoiter specific targets. Even more exotic nano- and picosatellites are on Israeli drawing boards. Israel is also working toward the capability of linking groups of satellites together in order to complement each other on specific reconnaissance assignments. It is no wonder that at least one American defense official commented that Israel has defense capabilities in space that are not possessed at the moment by the United States—though the nature of those capabilities was not disclosed in public.[23]

Lastly, while Israel officially adheres to the prohibition against deploying or employing weapons in space, the IAF might soon possess at least a rudimentary ability to shoot down hostile satellites. It has been suggested that the upgraded Arrow interceptor might be able to strike such targets in orbit. Moreover, it also might be possible to modify the new Silver Sparrow ballistic missile simulator to hit satellites.

Air Defense, Satellite Reconnaissance, and Israel's Strategic Landscape

Just as the IAF's UAV force has the potential to alter Israel's strategic landscape, so do the air force's reconnaissance satellites and IADS. At the upper end of the warfare spectrum, Israel may well face in-the-not-too-distant future the threat of an Iranian ballistic missile fleet armed with nuclear warheads. And it already faces the threat of Iranian and Syrian ballistic missiles armed with chemical and, perhaps, biological warheads. The combination of reconnaissance satellites and the Arrow interceptor system offers the prospect of bolstering Israel's deterrence against these weapons of mass destruction. Even if a prospective aggressor remained unconvinced that satellite (and, for that matter, UAV) surveillance coupled with the Arrow system could hermetically seal Israeli airspace—that is, could prevent every ballistic missile fired at Israel from reaching its target—it would still have to be concerned that this combination could at least defend effectively strategic point targets, such as air bases, naval facilities, and ballistic missile launchers, thus permitting the latter to respond in the form of a devastating "second strike," with its reputed arsenal of air-, sea-, and land-based nuclear weapons. Neither Iran nor Syria, most likely, would be very quick to employ weapons of mass destruction against Israel if it believed that the combination of the latter's reconnaissance systems, including satellites, and IADS rendered its second strike capability essentially invulnerable to a disarming "first strike."

At the lower end of the warfare spectrum, though systems like David's Sling and Iron Dome might not constitute a foolproof deterrent against rocket bombardment by Hizbullah or Hamas, when combined with the IDF's offensive power, they may well induce these organizations to think much longer and harder about engaging Israel in an asymmetrical war. Short- and medium-range rocket fire aimed at population and industrial centers, after all, is the main threat these terrorist organizations pose to Israel. If they believe that the Israeli IADS can largely defend those centers, they may well conclude that they cannot inflict enough death and destruction on Israel in order to justify the damage that they would be forced to absorb in return. In the event that deterrence nevertheless fails, of course, David's Sling and Iron Dome would be able to substantially mitigate the death and destruction inflicted on the Israeli home front (if the past successes of the latter system are any guide to the future), thereby giving the Israeli government and the IDF more freedom to smash the military capabilities of Hizbullah and/or Hamas.

Conclusion

The IAF possesses impressive ground-based air defense and space-based reconnaissance capabilities that are set to grow further in the future. From very humble beginnings in the 1950s and '60s, the IAF's IADS is now not only able to defend against aircraft, but also against ballistic missiles and rockets. Moreover, whereas this IADS formerly had no real ability to defend Israel's civilian population and industry, it now can do so to a considerable extent. Similarly, whereas Israel had no space-based reconnaissance assets a mere quarter-of-a-century ago, it now boasts four active intelligence satellites, including one that fields among the most advanced SAR systems in the world. Any potential aggressor must reckon with these formidable IAF capabilities before deciding to engage Israel in conflict.

6

The Infrastructure of the Israel Air Force: Bases and Squadrons

An air force cannot fulfill its roles in defense of the state without a dedicated infrastructure on the ground to support it, and the Israel Air Force (IAF) is certainly no exception in this regard. From its rudimentary origins in the 1947–49 War of Independence, the IAF's infrastructure has developed over the past six decades into a sprawling and sophisticated complex of command and control (C^2) facilities, air traffic control and training installations, and, of course, air bases. Indeed, it is safe to conclude that the IAF's infrastructure today is among the most advanced and efficient in the world.

IAF headquarters is ensconced in a modern office tower inside the Kirya, the sprawling military compound in central Tel Aviv that also houses the headquarters of the Israel Defense Forces (IDF) and the Defense Ministry. In addition to this skyscraper, the air force also has a heavily fortified underground command center somewhere inside the Kirya.[1] Additional fortified undergound command centers apparently exist in the Negev and Galilee, though their precise locations are not known to the public. One or both of these centers would presumably be activated in case the main center located inside the Kirya were to be rendered inoperative as a result of attack.

The IAF, which has ultimate control of the State of Israel's airspace, possesses an extensive air traffic control system, with installations located throughout the state. The main air traffic control center sits atop Mount Meron in the Galilee, the highest point within Israel proper. From this facility—and others like it—every aircraft flying to, from, and within Israeli airspace is closely monitored for any sign of trouble. Additionally, the IAF has integrated bird-watching radar into its air traffic control system, because Israel serves as a corridor for hundreds of millions of birds migrating between Europe and Africa. The inclusion of this technology has drastically cut the number of bird strikes, not only sparing the lives and limbs of IAF pilots, but also avoiding the loss of valuable aircraft.[2]

The IAF also possesses modern training ranges, with all of the associated instrumentation, in the open spaces of the Negev. The air force's fighter-bombers and attack helicopters use these ranges to prepare for air-to-air combat as well as air-to-ground attacks, particularly those involving the suppression of surface-to-air (SAM) and anti-aircraft artillery (AAA) batteries.[3] Still, Israel's small size limits the amount of realistic training that can be done in home skies. Historically, the IAF conducted much of its training over the Mediterranean Sea or, during the period from 1967–82, over the wide expanses of the Sinai. During the past couple of decades, however, the increasingly demanding missions assigned to the IAF has necessitated the search for overseas training grounds in order to familiarize pilots with the different kinds of terrain over which they may be called to fly in the future. In the 1990s, the air force took advantage of Turkish airspace for training purposes; however, as Turkish–Israeli relations have steadily soured over the past decade, the IAF has turned to the airspace of such states as Italy, Greece, Romania, Cyprus, and Bulgaria.[4]

Air Bases and Squadrons

To construct sophisticated headquarters facilities, air traffic control systems, and training ranges, of course, would be pointless if a sizable air force did not exist to exploit them. And a sizable air force, in turn, requires an extensive infrastructure of air bases. For a state of its geographical size, Israel has a rather impressive array of air bases. The IAF boasts eight major airfields—Ramat David, which is located in the lower Galilee; Tel Nof, Hatzor, and Palmachim, which are located on the coastal plain south and southeast of Tel Aviv; and Nevatim, Hatzerim, Ramon, and Uvda, which are spread throughout the Negev. The IAF also has at its disposal some smaller airstrips, such as Sde Dov in north Tel Aviv and Megiddo just south of Ramat David, as well as forward helicopter bases, such as Biranit in the northern Galilee. In a pinch, the IAF could also operate out of Ben-Gurion Airport, Israel's primary civilian airfield, which it actually did until a few years ago. Some of Israel's major highways are able to serve as emergency runways as well.

Very little of this infrastructure was in place on the eve of Israel's establishment in mid-1948. In the War of Independence, the IAF flew out of both abandoned British air bases and hastily improvised airstrips.

In either case, these installations could best be described as primitive, with not much in the way of proper storage and maintenance areas and procedures. Significant growth in the IAF's bases would not take place until after the war. During the 1950s and '60s, the air force's principal bases—Ramat David, Tel Nof, and Hatzor—were enlarged and modernized ex-Royal Air Force (RAF) installations. The fighter-bombers and helicopters that waged the 1956 Sinai Campaign and the 1967 Six-Day War flew from these airfields.[5] The IAF's transport aircraft operated out of the military section of Ben-Gurion Airport (then known as Lod Airport). Israel constructed its first air base, Hatzerim, in the northern Negev during the mid-1960s, in part to serve as the home of the air force's flight academy. These four bases, plus the military section of Ben-Gurion Airport, housed the IAF during the 1969–70 War of Attrition and the 1973 Yom Kippur War, though the air force also had at its disposal captured Egyptian airfields and improvised airstrips throughout the Sinai.

Currently, Ramat David hosts three F-16 Fighting Falcon squadrons, two of which are tasked primarily with the air superiority role and one of which is tasked primarily with the close air support (CAS) and interdiction roles, especially as they pertain to the suppression of an integrated air defense system (IADS).[6] 110 "Knights of the North" Squadron and 117 "The First Jet" Squadron, both of which fly single-seat F-16Cs, specialize in the air superiority role. Indeed, 117 squadron, one of the most illustrious in the IAF, has chalked up quite an impressive war record, including more than 120 victories in air combat.[7] 109 "The Valley" Squadron operates two-seat F-16Ds, which are fitted out with all manner of ultrasophisticated electronic systems and precision-guided munitions (PGMs). Ramat David also houses the IAF's sole naval helicopter unit, 193 "Defenders of the West" Squadron, which is equipped with the AS565MA, when these machines are not flying off the decks of the Israel Navy's Sa'ar 5 missile corvettes. Due to its close proximity to hostile territory, the base suffered damage from Jordanian artillery fire in the Six-Day War and Syrian rocket fire in the Yom Kippur War; however, flight operations were not impeded in either conflict.

Hatzor, in a sense, is a smaller version of Ramat David. It houses two F-16 squadrons, one in the air superiority role and one in the CAS and interdiction roles. 101 "The First Fighter" Squadron, perhaps the most famous in the IAF, flies the F-16C, like 110 and 117 squadrons at Ramat David. This squadron, too, has achieved many air combat victories over the course of its existence. 105 "The Scorpion" Squadron is a sister unit of 109 squadron. It, too, flies the F-16D.

Tel Nof and Hatzerim, on the other hand, are considerably larger bases. Tel Nof houses two F-15 Eagle units, 133 "Knights of the Twin Tail" Squadron and 106 "The Point of the Spear" Squadron, whose principal task is air superiority, as well as two CH-53 Sea Stallion helicopter units, 114 "The Night Leaders" Squadron and 118 "The Nocturnal Birds of Prey" Squadron, whose principal tasks are transport, including the insertion and extraction of special forces units deep inside hostile territory, and casualty evacuation. 133 "Knights of the Twin Tail" Squadron has achieved more than 50 air combat victories in the F-15 without suffering a single battle loss.[8] The CH-53 squadrons, as their names indicate, are particularly active during the night.[9]

Tel Nof is also home to at least one unmanned aerial vehicle (UAV) unit, 210 "The Eitan UAV" Squadron, which flies the Heron TP drone, as well as the personnel of Unit 669, the IAF's aeromedical rescue and evacuation unit. Depot 22, the air force's central maintenance facility—which overhauls and repairs aircraft and helicopters, investigates aerial accidents, and solves hardware-related engineering problems—is located at Tel Nof, as is the IAF Flight Test Center, known by the Hebrew acronym MANAT, which evaluates aircraft, weapons systems, and electronic systems. The IAF, incidentally, is one of the few air forces in the world with its own flight test center.[10] The center has its own aircraft, but sometimes borrows additional machines from the IAF's fighter-bomber squadrons. Finally, Tel Nof is home to the IDF's parachute training school.

Like Tel Nof, Hatzerim hosts two fighter-bomber squadrons. 69 "The Hammers" Squadron flies the F-15I, a newer long-range strike version of the F-15, and 107 "Knights of the Orange Tail" Squadron flies the F-16I, a newer long-range strike version of the F-16. The primary role of both of these fighter-bomber squadrons is to conduct strategic attacks. 69 Squadron most likely carried out the attack on Syria's nuclear reactor in 2007. 123 "The Southern Bells" Squadron today operates the UH-60 Blackhawk, and its principal tasks are transport and casualty evacuation. Hatzerim is also home to the IAF's Flight Academy, as well as the air force's small aerobatic team. 102 "The Flying Tigers" Squadron, which still flies the old A-4 Skyhawk (soon to be replaced by a more modern Italian aircraft), trains flight academy graduates for deployment to F-15 and F-16 squadrons.[11]

From the mid-1970s to the mid-1980s, the IAF's infrastructure underwent rapid development. Three new airfields were constructed in the Negev—Nevatim in the northern part of the desert, Ramon in the central part, and Uvda at the southern tip—to replace the air bases abandoned when Israel returned the Sinai to Egypt as part of the peace agreement between the two states. Earlier, during the mid-1970s, the IAF greatly expanded Palmachim, which had long served as a missile testing and space launching installation, eventually turning it into the air force's principal helicopter and UAV base. Presently, Palmachim is home to 160 "The Northern Attack" Squadron, an AH-1 Cobra attack helicopter unit, and 124 "The Rolling Sword" Squadron, a UH-60 unit. The former is primarily employed in the CAS and interdiction roles, while the latter engages in transport and casualty evacuation. At least two UAV units, 200 "The First UAV" Squadron, which flies the Heron drone, and 166 "The UAV" Squadron, which flies the Hermes 450 drone, operate out of Palmachim, which also hosts one of the IAF's Arrow anti-ballistic missile interceptor batteries. The Shaldag unit is quartered on this base as well.

Initially one of the IAF's less important bases, Nevatim has nowadays grown into one of the most important air force installations. It houses two F-16 units, 116 "Defenders of the South" Squadron and 140 "The Golden Eagle" Squadron, whose primary tasks are air superiority, particularly in regard to the airspace around Israel's nuclear reactor complex at Dimona, and operational training of young pilots directly out of the flight academy. Both of these squadrons fly among the oldest F-16s in the IAF fleet. In recent years, all of the IAF squadrons formerly flying out of the military section of Ben-Gurion Airport have been relocated to Nevatim. 120 "The International" Squadrron operates a mix of Boeing 707 aircraft as tankers and electronic warfare platforms, as well as the 1124 Westwind aircraft in the naval reconnaissance role. Two C-130 Hercules units, 103 "The Elephants" Squadron and 131 "The Yellow Birds" Squadron, fill transport and air-to-air refueling roles.[12] An intelligence-gathering and electronic warfare unit, 122 "The Dakota" Squadron, which operates two distinct versions of the Gulfstream jet, is also housed at Nevatim. These aircraft are perhaps the most advanced of their types in the world.

Ramon Air Force Base boasts three F-16I units, the remainder of the IAF's long-range strategic attack force—119 "The Bat" Squadron, 201 "The One" Squadron, and 253 "The Negev" Squadron. The air force's two AH-64 Apache attack helicopter units, 113 "The Hornet" Squadron and 190 "The Magic Touch" Squadron, also call Ramon home. The IAF's "aggressor squadron," which helps to train other air force squadrons in air-to-air and air-to-ground combat, is located at Uvda Air Force Base, located just to the north of Israel's Red Sea resort town of Eilat. 115 "The Flying Dragon" Squadron is unique in the IAF in that it consists of both F-16 fighter-bombers and AH-1 attack helicopters. The extremely dry climate in the vicinity of Uvda probably accounts for the fact that the IAF stores most of its older, out-of-service aircraft here.[13]

Not a major air base by any stretch of the imagination—it cannot even handle jet aircraft—Sde Dov in north Tel Aviv deserves a mention in any review of IAF airfields, as it houses two intelligence-gathering and liason units, 100 "The Flying Camel" Squadron and 135 "The Kings of the Air" Squadron, both of which fly a mix of small, propeller-driven aircraft. In the former role, these aircraft have been instrumental in collecting and disseminating information for targeted attacks, especially in Gaza.[14] Sde Dov may be shut down sometime in the near future so that the prime real estate on which it sits can be developed for civilian purposes. If this development comes about, these squadrons would most likely be relocated to Palmachim.

With the prominent exceptions of Tel Nof in the War of Independence and Ramat David in the Six-Day War and the Yom Kippur War, the IAF's air bases have never come under attack. Nevertheless, they are heavily fortified, as the air force expects that they will be targets of medium- and long-range missiles and rockets in a future conflict. According to a recent IAF commander:

> The air force assumes that it will have to operate while its bases are under missile attack. The implications of this for the bases are enormous, and it mainly boils down to training and educating the personnel to continue functioning under attack. It also requires protecting the work places so that operations can continue. We have prepared for this in many ways.[15]

Each of the major air bases has enough hardened air shelters (HAS) to accommodate all of the aircraft deployed to it. Some of these bases may also have underground aircraft shelters, though there is no information in the public domain to support this speculation. Fuel and ordnance stocks, on the other hand, are almost certainly kept in underground bunkers when not being readied for immediate use. All of the bases make extensive use of camouflage. Each base has a passive defense officer whose function is to ensure that flight operations continue unimpeded should the installation come under attack.[16] Repair teams and procedures are in place to ensure that cratered runways would be back in service in short order. Each base also has active defenses, including SAM and AAA protection. Perimeter defense in the unlikely event of some sort of ground assault is provided by fencing backed up by foot- and/or vehicleborne patrols.

Though not an air base per se, mention should be made of the IAF's technical school in Haifa. At this facility, the air force trains technicians in all of the trades—for example, aircraft mechanic, flight controller, and radar operator—that are necessary to sustain a modern air force. Trainees are given hands-on instruction in the design and maintenance of aircraft, electronic systems, and so on. The IAF also sponsors a number of technical high schools throughout Israel that prepare young men and women for induction into the air force by giving them both a general and technical education. And a joint IAF–Israel Air Force Center (IAFC) program helps to promote leadership qualities in future IAF personnel through the latter's National Center for Leadership and Character Development.[17]

Conclusion

Israel has built up a large air force over the past six plus decades in order to defend its national security interests. Concomitantly, it has created an extensive network of air bases, C^2 facilites, air traffic control installations, training ranges, and technical schools to guarantee that the IAF's aircraft can operate at maximum efficiency. With no end in sight to the ongoing Arab–Israeli conflict, Israel will undoubtedly continue to invest considerable human and matériel resources in the air force's infrastructure.

Conclusion: The Past and Future Contributions of Airpower to Israeli National Security

The State of Israel has thrived over the past 60 plus years. Through a combination of massive immigration and natural growth, its population has swelled in leaps and bounds, from six hundred thousand souls in 1948 to well over seven million today. Israel's economy has witnessed an even more spectacular rate of growth. A tiny, mainly agricultural economy, with few natural resources at its disposal, has evolved into a technological powerhouse that exports state-of-the-art goods and services worldwide in fields as diverse as biotechnology, Internet technology, and aerospace technology. Progress in other areas—for example, health and education—has been similarly great. Furthermore, Israel's commitment to democratic norms and humanitarian values, always strong, has deepened over time. Perhaps no other country in the post-Second World War era, in short, can boast of so many accomplishments.

All of this progress has occurred in the midst of an incessantly hostile and tumultuous environment, one in which Israel's Arab opponents—with the unabashed approval of much of the international community and the unconcerned acquiescence of much of the rest of it—have sought to terminate its very existence. Coupled with its willingness and ability to fight, a sensible national security doctrine—one that has taken account of its territorial dimensions, its demographic and economic limitations, its lack of formal alliance partners, among numerous other variables—has permitted Israel to weather this long-standing storm quite nicely, occasional setbacks notwithstanding.

Airpower has been an important element in this national security doctrine, at least since the aftermath of the 1956 Sinai Campaign, when it became a central component in the joint air–land mobile warfare concept embraced by the Israel Defense Forces (IDF). Since the 1967 Six-Day War, the Israel Air Force (IAF) has had a central part in every type of conflict waged by the IDF, including high-intensity interstate maneuver wars, high-intensity interstate and asymmetrical attrition wars, low-intensity counterinsurgency campaigns, and special operations warfare.

Though by no means a panacea—let alone decisive in its own right—airpower has generally served Israel well in its various conflicts. The IAF has maintained air supremacy over the home front at all times, not only sparing the state's population centers and industrial assets from major damage, but also ensuring that the IDF's land forces, always heavily dependent on reserve manpower, have been able to mobilize and deploy to the fronts in a timely manner. Furthermore, even if the IAF has not always been able to secure air superiority over the battlefield in the face of Arab integrated air defense systems (IADSs), it has always retained air superiority vis-à-vis Arab air forces, thereby ensuring that IDF land forces have never come under effective and sustained air attack.

Militarily speaking, the IAF's strategic attacks—whether against leadership personnel, long- and medium-range rocket launchers, command and control (C^2) facilities, nuclear weapons plants, supply depots, and the like—have generally been successful. Air strikes against leadership personnel and rocket launchers, for example, have undoubtedly prevented much death and destruction within Israel, and they have derailed the nuclear plans of two of the state's most implacable opponents. Blame for the fact that these strikes have sometimes not resulted in the political outcomes desired by Israel cannot be laid at the IAF's feet. The problem has been one of ends–means conceptualization on the part of the state's politico-military leadership, not execution on the part of its airmen.

The record with respect to interdiction and close air support (CAS) is less clear. On the one hand, the IAF has engaged in some interdiction and CAS efforts that have had a significant impact on the IDF's war efforts—for example, on the Golan front in the opening days of the 1973 Yom Kippur War. On the other hand, the air force has sometimes not possessed the assets (appropriate aircraft and ordnance, real-time intelligence, proper C^2 facilities), or sometimes not had sufficient control of the airspace above the battlefield, to support the land fighting to any great extent—for example, on the Sinai front in the opening days of the Yom Kippur War.

The air force has functioned efficiently in its ancillary roles. Whether in high-intensity maneuver or attrition wars, low-intensity counterinsurgency campaigns, or special operations warfare, it has always been able to airlift troops to the battlefield or deep behind the front lines.

Witness the heliborne raids on the Egyptian interior during the 1969–70 War of Attrition or the 1976 hostage-rescue raid on Entebbe airport. The IAF has routinely medevaced wounded troops from the battlefield, saving many lives in the process, and it has just as routinely airlifted supplies to land forces in the field.

The picture in regard to airborne reconnaissance activities is more complex. The IAF has always had good reconnaissance capabilities; however, before the advent of unmanned aerial vehicles (UAVs) able to furnish real-time information, the intelligence gathered by its aircraft typically could not be disseminated and acted upon in an expedient manner. The employment of UAVs during the 1982 Lebanon War fundamentally changed this situation. In the conflicts that Israel has fought since the Lebanon War, especially its asymmetrical wars with Hizbullah and Hamas, as well as its counterinsurgency campaign during the second intifada (or the so-called al-Aqsa Intifada), UAVs have consistently provided real-time information to both air and land commanders such that their assets have been able to strike targets with a high degree of accuracy in very short order. The "sensor-to-shooter" loop has nowadays been reduced in certain cases to a matter of mere seconds.

With all of the attention paid to Arab IADSs, especially during the Yom Kippur War, the fact that Israel has had an effective IADS of its own has not received much fanfare; however, this system proved itself during that war by knocking down scores of Arab aircraft. Over the past two decades, the IAF has supplemented and upgraded its IADS with interceptor systems capable of destroying ballistic missiles and short-range rockets. Indeed, its Iron Dome interceptor system has already shown itself quite capable of destroying short-range rockets. And, during the same time frame, the IAF has supplemented its extensive airborne reconnaissance assets with highly sophisticated space-based satellites.

Finally, it must be mentioned that the IAF's participation in humanitarian operations around the globe has contributed—albeit to a rather limited extent—to Israeli national security by generating a measure of goodwill toward the state from quarters not generally sympathetic to it.

So much for the IAF's past contribution to Israeli national security. How might it assist in defending the state moving forward? What combat and "noncombat" roles will it perform in the future?

Into the Future

For the foreseeable future, the IAF will continue to fulfill the same roles as in the past. Air dominance in the form of the attainment of air supremacy over the home front and air superiority over the battlefield will be its paramount priority. Air supremacy over the home front seems all but assured in any future conflict, as the IAF's excellent air-to-air capabilities, coupled with the poor quality of its most likely opponents' air forces, suggests that their aircraft (or drones) would be unable to penetrate Israeli airspace in meaningful numbers, even if they were not destroyed during air base attacks. Air superiority over the battlefield should be more complete than in the past, as the IAF now has the capabilities to overcome any opponent's IADS through a combination of cyber/electronic warfare, not to mention "kinetic" attack. Indeed, the precision-guided munitions (PGMs) that arm today's IAF aircraft can destroy both surface-to-air (SAM) and anti-aircraft artillery (AAA) batteries at stand-off range so long as real-time intelligence on the whereabouts of these positions is available.[1]

The strategic attack role will take on even greater urgency than in the past, especially in light of the proliferation of ballistic missiles and long-, medium-, and short-range rockets in the hands of Israel's opponents near and far. The air force's plan to acquire a small fleet of F-35 Lightning II fifth-generation stealth fighter-bombers, whose primary role is strategic attack, to supplement its F-15 Eagle and F-16 Fighting Falcon strike aircraft is a clear indication of this trend.[2] The IAF's performance in the 2006 Second Lebanon War clearly displayed its capability to knock out long- and medium-range rockets, a capability that has been considerably enhanced over the past few years. Short-range rockets and ballistic missiles are another story. The almost certain inability to eliminate entirely these threats solely from the air in any future conflict has convinced Israel to strengthen substantially its ground-based air defenses. In addition to the already deployed Arrow anti-ballistic missile and the Iron Dome anti-short-range rocket interceptor systems, the IAF will also field the David's Sling interceptor system—optimized for use against medium-range rockets, cruise missiles, and aircraft—in the coming years. It is quite likely, too, that the IAF will be called upon to mount strategic attacks against targets ranging from leadership personnel to weapons-of-mass-destruction facilities in any future conflict.

Regardless of the form of any future conflict—high-intensity maneuver war, high-intensity attrition war, or counterinsurgency campaign—the IAF will be expected to engage in vigorous interdiction and CAS efforts on behalf of IDF land forces. The likely achievement of air superiority over the battlefield, coupled with the acquisition of all manner of PGMs and C⁴ISR (command, control, communications, computers, intelligence, surveillance, and reconnaissance) technology, should make these efforts more effective than in the past. Though the IAF has never relished the CAS role, it has now committed itself more firmly than in the past to closer cooperation with IDF land forces, particularly through the integration of air control officers into these forces down to brigade level.[3] The air force has also created a joint air–land "cooperation center" under its aegis, graduating its first crop of "students" in mid-2012. This center trains IDF land forces officers to utilize IAF assets to assist those forces.[4] Helicopter gunships and UAVs will likely play a central part in any future interdiction and CAS efforts.

The IAF will continue to fulfill its traditional ancillary roles as well. Troop transport, logistical support, and medevac will remain priorities. Troop transport will undoubtedly remain the preserve of manned aircraft, as the IDF will be loathe to entrust the welfare of large numbers of its troops to UAVs. For psychological reasons, it is necessary to keep the human element "in the loop" when the lives of large numbers of men and women might be on the line. Logistical support and medevac, however, are roles that could well be taken over largely (if not completely) by drones in the future. Israel is already working on a medevac UAV that will be able to remove the wounded from a "hot" landing zone more quickly and more safely than manned helicopters or transport aircraft. Reconnaissance, too, will more and more become the preserve of drones, especially as their payloads, ranges, and time over target increase. Their loitering capability over a specified area—a capability not possessed by manned aircraft—makes them indispensable for providing real-time "actionable" information. The IAF will also incorporate to a greater extent imagery from Israel's large (and growing) contingent of satellites into its reconnaissance assets.

Finally, given Israel's genuine humanitarian impulse, as well as its desire to make friends around the world, the IAF is likely to be called upon to engage in relief operations on an ever larger scale in the years ahead.

A recent IAF commander has nicely captured the air force's past and future centrality to the protection of Israel.

> Our security concept stands on four legs: maintaining a deterrent force, the ability to achieve a military decision, intelligence warning, and defense—and in each of the four areas[,] the air

> force plays a central role whose importance will not lessen in the coming years.
>
> If we examine each of the parameters, then it [is] clear the air force has played a leading part in the deterrence that [has] enabled Israel to flourish for decades. If deterrence fails[,] then we turn to our ability to achieve victory, which is based on . . . activity in enemy territory as much as possible—and not in our territory, [which is] limited in size and lacking in depth. Here too the air force's importance is invaluable[, because] it alone possesses the [capabilities] to operate at different ranges in different places. Today[,] there is no other means or force or branch that can deliver precision firepower at as [high an output] and in as short a time as the air force. The air force alone can move from theater to theater in a matter of hours. . . .
>
> As far as early intelligence warning goes, the air force has a central part to play as the operator of various means of . . . reconnaissance. And in the area of defense—not just active defense against missiles, whose current development is moving forward at full steam—but defense against enemy aircraft, defense of the nation's skies in the classic sense of "defense"—this has been and remains our historical purpose. In general, we can say that all of the main parameters in Israel's security concept come under one umbrella—the air force.[5]

All in all, as a result of its impressive capabilities and proven track record, as well as the nature of the threats faced by Israel, the IAF seems set to shoulder an even greater part of the state's overall defense burden than in the past. The air force alone, after all, has the ability to defend the state from long-range threats (e.g., Iran's ballistic missile fleet). And many threats closer to home cannot be effectively countered without a major air force effort in battle. One would hope, of course, that the IAF never has to go into action again; however, in the tumultuous Middle East, this prospect is surely nothing more than wishful thinking.

Appendix: The Historical Evolution of the Israel Air Force's Fighter-Bomber Inventory

The Israel Air Force (IAF) has passed through a number of distinct periods in respect to its inventory of fighter-bombers, the core of its fighting power. From the late 1940s through the early 1950s, it acquired a hodgepodge of machines from whatever sources displayed a willingness to furnish the State of Israel with arms. From the mid-1950s through the mid-1960s, the air force built itself around French aircraft. From the late 1960s through the mid-1980s, the IAF largely re-equipped itself with American and Israeli machines. And, from the late 1980s forward, the air force has acquired American aircraft, but with a very pronounced "made-in-Israel" component.

Ironically enough, the Czechoslovakian-built German Bf-109, known as the Avia S-199, sold to Israel by Czechoslovakia with the tacit permission of the Soviet Union, was the first fighter-bomber to enter IAF service. The British Spitfire, also purchased from Czechoslovakia, arrived shortly thereafter. These two aircraft made up the bulk of the IAF's fighter-bomber inventory during the 1947–49 War of Independence, though at least one American P-51 Mustang, three American B-17 Flying Fortresses (a dedicated long-range heavy bomber), and a few other combat aircraft found their way into Israeli hands. In the years immediately following the war, the IAF continued to build up its fleet of fighter-bombers with Spitfires, Mustangs, and British Mosquitos.

The jet age opened for the air force in the early 1950s, when it acquired the British Meteor; however, by the mid-1950s, the IAF had entered the "French era," with the receipt of the Ouragan and Mystère IV, both of which saw action in the 1956 Sinai Campaign. In the aftermath of that conflict, the air force added the Vautour, the Super Mystère, and, finally, the Mirage III. French aircraft equipped all of the IAF's fighter-bomber squadrons during the 1967 Six-Day War, a conflict that earned the Mirage III in particular a worldwide reputation for excellence.

Despite its positive experience with French aircraft in the Sinai Campaign and the Six-Day War, the IAF had long shown a strong interest in American hardware. As the United States and Israel drew closer together during the 1960s, the air force entered the "American era," when it acquired the A-4 Skyhawk and the F-4 Phantom.[1] These fighter-bombers played a significant part in Israeli air operations during the 1969–70 War of Attrition, taking the place of the IAF's obsolescent Ouragans, Mystère IVs, and Vautours. In the early 1970s, the air force also added the Nesher, a home-built version of the French Mirage V fitted with an American engine, to its fighter-bomber inventory. By the time of the 1973 Yom Kippur War, the IAF's fighter-bomber squadrons—except for three Mirage III/Nesher and one Super Mystère squadrons—were equipped entirely with Skyhawks and Phantoms.

In the wake of this war, the IAF upgraded its fighter-bomber squadrons with more advanced models of the Skyhawk and Phantom. Simultaneously, it fielded the indigenously produced Kfir, which would eventually replace the remaining Mirage IIIs/Neshers in air force service. In the mid-1970s, the IAF acquired its first squadron of American F-15 Eagles; and, by the early 1980s, it added the American F-16 Fighting Falcon to the ranks of its fighter-bomber squadrons. All of these types participated heavily in the 1982 Lebanon War.

Despite a steady supply of advanced American aircraft, however, Israel sought to design and build another fighter-bomber of its own. Unlike the Nesher and the Kfir, which represented successively improved versions of the French Mirage V, the Lavi was intended to be a state-of-the-art aircraft that would become the IAF's principal strike vehicle. In the end, the Lavi never got beyond the prototype stage, because of American opposition to its development.[2]

The demise of the Lavi, whose capabilities would have been roughly equivalent to those of the F-16I, gave rise to yet another era in IAF fighter-bomber procurement, what might be termed the "American–Israeli" period of hybrid aircraft. The platforms—F-15s, F-16s, and, in the future, F-35 Lightning IIs—remain American, but much of the avionics and ordnance are Israeli in origin. The IAF, of course, has a long history of incorporating Israeli-made avionics and ordnance into foreign-built aircraft—the Shafrir II air-to-air missile on the Mirage III, for example—however, this trend has mushroomed since the late 1980s. Nowadays, most of the electronic warfare systems on the IAF's F-15s and F-16s are homegrown. So are the principal air-to-air (Python IV, Python V, and Derby) and air-to-surface (Popeye and Delilah) missiles. IAF fighter-bombers still employ much in the way of American avionics and ordnance, but the trend toward making these aircraft more and more "blue and white" is unmistakable. The F-35, when all is said and done, will probably become the most heavily modified aircraft of all in IAF service.

Notes

Introduction (Part I): Israeli National Security and Airpower

1 General, though now partially dated, treatments of Israel's national security doctrine may be found in Yoav Ben-Horin and Barry Posen, *Israel's Strategic Doctrine* (Santa Monica, CA: The Rand Corporation, 1981); Eliot A. Cohen, Michael J. Eisenstadt, and Andrew J. Bacevich, *Knives, Tanks, and Missiles: Israel's Security Revolution* (Washington, D.C.: The Washington Institute for Near East Policy, 1998); Michael Handel, *Israel's Political–Military Doctrine* (Cambridge: Harvard University Press, 1973); Michael Handel, "The Evolution of Israeli Strategy: The Psychology of Insecurity and the Quest for Absolute Security," pp. 534–578 in Williamson Murray, MacGregor Knox, and Alvin Bernstein (eds.), *The Making of Strategy: Rulers, States, and War* (New York: Cambridge University Press, 1994); Ariel Levite, *Offense and Defense in Israeli Military Doctrine* (Boulder, CO: Westview Press, 1989); Bard E. O'Neill, "Israel," pp. 497–541 in Douglas J. Murray and Paul R. Viotti (eds.), *The Defense Policies of Nations: A Comparative Study*, Third Edition (Baltimore: The Johns Hopkins University Press, 1994); David Rodman, "Israel's National Security Doctrine: An Introductory Overview," *MERIA Journal*, Vol. 5, No. 3 (September 2001), pp. 71–86; David Rodman, "Israel's National Security Doctrine: An Appraisal of the Past and a Vision of the Future," *Israel Affairs*, Vol. 9, No. 4 (Summer 2003), pp. 115–140; Israel Tal, *National Security: The Israeli Experience* (Westport, CT: Praeger, 2000); and Avner Yaniv, *Deterrence Without the Bomb: The Politics of Israeli Strategy* (Lexington, MA: Lexington Books, 1987).

2 The fundamental distinction between preventive and preemptive war concerns urgency. A preventive war is fought to impede a latent threat from developing into a manifest threat. A preemptive war, on the other hand, is fought to counteract a manifest threat.

3 Hizbullah, an indigenous Lebanese Islamist terrorist organization created by Iran in the early 1980s and nurtured by it ever since, filled the vacuum left by the expulsion of the Palestine Liberation Organization (PLO) from Lebanon. By the mid-1980s, it had become

Israel's principal nemesis in Lebanon, and it remains an implacable foe of the state to the present day. Hamas is a Palestinian Islamist terrorist organization that has carried out hundreds of homicide bombings and other attacks against Israelis from the early 1990s onward, not to mention thousands of rocket and mortar strikes against southern Israel from Gaza since the Israeli withdrawal in 2005. It is the Palestinian branch of the international Muslim Brotherhood.

4 For the IDF's current weapons inventory, see the military database at the Institute for National Security Studies at Tel Aviv University. This database in available on the Internet at *www.inss.org.il.*

5 Maneuver is a mode of warfare based upon rapid movement. It is particularly suitable for armies that wish to engage in swift offensive thrusts. Attrition, in contrast, is a mode of warfare that emphasizes firepower. It is particularly suitable for armies that do not wish—or are not capable of—swift offensive thrusts. Though the military doctrines of all states, in actuality, are built upon a combination of mobility and firepower, one usually takes precedence over the other.

6 The operational level of warfare refers to the way the military establishment of a state arranges and employs its combat branches, such as its air force. The tactical level of warfare refers to the way the military establishment arranges and employs its lower-level combat units, such as its air force squadrons.

Introduction (Part II): Israeli Airpower and the Arab–Israeli Conflict

1 In the early years of the twenty-first century, the Israel Air Force officially changed its name to the Israel Air and Space Force (IASF) in order to reflect the fact that it now operates in space as well. Nevertheless, this monograph shall refer to the IAF throughout, as most of the historical events described and analyzed in the following chapters occurred before the adoption of the new name.

2 On the Israeli–Soviet clash, see Benjamin S. Lambeth, *Moscow's Lessons from the 1982 Lebanon War* (Santa Monica, CA: The Rand Corporation, 1984), p. 28.

3 Much has been written about the IAF's history, particularly through the early 1980s. Among the more valuable general histories are Eliezer Cohen, *Israel's Best Defense: The First Full Story of the Israeli Air Force* (New York: Orion Books, 1993); Anthony H. Cordesman and Abraham R. Wagner, *The Lessons of Modern War (Vol. 1): The Arab–Israeli Conflicts, 1973–1989* (Boulder, CO: Westview Press, 1989); Trevor N. Dupuy, *Elusive Victory: The Arab–Israeli Wars, 1947–1974* (New York: Random House, 1978); Trevor N. Dupuy and Paul Martell, *Flawed Victory: The Arab–Israeli Conflict and the 1982 War in Lebanon* (Fairfax, VA: Hero Books, 1986); Chaim Herzog, *The*

Arab–Israeli Wars: War and Peace in the Middle East from the War of Independence Through Lebanon (New York: Random House, 1982); Robert Jackson, *The Israeli Air Force Story* (London: Tom Stacy, 1970); Edward N. Luttwak and Dan Horowitz, *The Israeli Army* (New York: Harper and Row, 1975); Bill Norton, *On the Edge: Aircraft of the Israel Air Force Since 1948* (Hinckley: Midland Publishing Ltd., 2002); Martin van Creveld, Steven L. Canby, and Kenneth S. Brower, *Air Power and Maneuver Warfare* (Maxwell Air Force Base, AL: Air University Press, 1994); and Ehud Yonay, *No Margin for Error: The Making of the Israeli Air Force* (New York: Pantheon Books, 1993).

4 See the appendix for a brief discussion of how the IAF's inventory of fighter-bombers has evolved over the past six plus decades.

5 France had been Israel's principal source of aircraft from the mid-1950s to the mid-1960s. On the eve of the Six-Day War, though, France slapped an arms embargo on Israel in order to curry favor with the Arab world, one that became permanent after the cessation of hostilities. The United States, which had already begun to supply Israel with tanks and aircraft prior to the Six-Day War, replaced France as Israel's chief source of arms. For the transition, see David Rodman, *Arms Transfers to Israel: The Strategic Logic Behind American Military Assistance* (Brighton and Portland, OR: Sussex Academic Press, 2007).

6 Islamic Jihad is a much smaller, but equally fanatical and bloody, Palestinian Islamist terrorist organization that sometimes competes with—and sometimes cooperates with—Hamas. Fatah, a "secular" Palestinian terrorist organization, was the main constituent group of the PLO, and it now has close links to the Palestinian Authority (PA), the internationally recognized leadership of the Palestinian population resident in Judea, Samaria, and Gaza.

7 The author would like to thank United States Air Force (USAF) Lieutenant Colonel Paul D. Berg for his insightful comments about some of the terminology appearing in this monograph. Responsibility for any nontraditional and/or imprecise use of this terminology, of course, lies solely with the author.

1 Airpower and Maneuver Warfare: The Israel Air Force in the 1967 and 1973 Wars

1 A meticulous descriptive account of the preparation and execution of Operation Focus can be found in Shlomo Aloni, *The June 1967 Six-Day War (Volume A): Operation Focus* (Bat Hefer: Isradecal, 2008). This account, by a well-known historian of Israeli airpower, relies on a combination of official IAF records and interviews with participating aircrews.

2 The standard air base attack sortie consisted of one bombing run to

crater runways, followed by three strafing passes, primarily to destroy aircraft, but also to smash other targets of opportunity (e.g., heavy equipment and maintenance workshops). E-mail communication, August 10, 2008, with Lieutenant Colonel (Ret.) Amos Cohen, who flew two air base attack sorties on the first day of the war.

3 Aloni, *The June 1967 Six-Day War*, pp. 179–180. In a June 28, 2011 conversation with the author at the Israel Air Force Center (IAFC), Herzliya, Israel, Ilan Hayit, a Mirage III pilot during the Six-Day War, related a telling anecdote in connection with the Egyptian air force's inability to recover quickly from Operation Focus. He had been one of the IAF pilots to crater the runways at Cairo's airport. Many years later, as an El Al (Israel's national airline) pilot, he distinctly remembered the bumpy landings at this airport, which testified to the Egyptians' inability to repair properly their damaged runways.

4 Ibid., pp. 179–180.

5 Ibid., pp. 54, 179.

6 These figures come from the IAF Historical Branch. They are taken from Kenneth M. Pollack, "Air Power in the Six-Day War," *The Journal of Strategic Studies*, Vol. 28, No. 3 (June 2005), pp. 471–503. For the figures, see the table on p. 478.

7 For a concise and trenchant examination of IAF CAS and interdiction attacks on each of the three fronts—the Sinai, the Judean and Samarian, and the Golan—see ibid., pp. 477–485. On the Sinai front, also see Martin van Creveld, "Israel: Maneuver Warfare, Air Power, and Logistics," in Martin van Creveld, Steven L. Canby, and Kenneth S. Brower, *Air Power and Maneuver Warfare* (Maxwell Air Force Base, AL: Air University Press, 1994), pp. 153–192. For the Sinai front, see pp. 168–169.

8 For the conclusion that the IAF destroyed few tanks (on all three fronts), see the report by the Joint Technical Coordinating Group for Munitions Effectiveness, *Special Report: Survey of Combat Damage to Tanks*, Volumes I–III (Washington, D.C.: Defense Intelligence Agency, 1970). Also see Pollack, "Air Power in the Six-Day War," pp. 483–485 and van Creveld, "Israel," pp. 168–169.

9 For one instance of a paratrooper drop behind Egyptian lines, see van Creveld, "Israel," p. 166.

10 Ibid., p. 168.

11 On American pressure to refrain from a preemptive strike, see the October 6, 1973 telegram from Secretary of State Henry Kissinger to President Richard Nixon in Middle East War Memos and Miscellaneous (October 1–October 17), National Security Council File, Nixon Presidential Materials Project, Box 664, United States National Archives, College Park, Maryland and the October 7, 1973

memorandum of conversation between Kissinger and Israeli Ambassador Simcha Dinitz in Records of Henry Kissinger, 1973–1977, Record Group 59, State Department Records, Box 25 (Category C 1973 Arab–Israeli War), United States National Archives, College Park, Maryland. Also see Michael Brecher, *Decisions in Crisis: Israel, 1967 and 1973* (Berkeley: University of California Press, 1980), pp. 177–178, 187–188, and 197–201.

12 In a June 25, 2007 conversation with the author at the IAFC, Brigadier General (Ret.) Itzhak Amitay, who flew many combat missions on the Sinai front during the war, confirmed that the IAF was making headway against the Egyptian IADS before the IDF high command scrubbed the operation. Also see Ehud Yonay, *No Margin for Error: The Making of the Israeli Air Force* (New York: Pantheon Books, 1993), pp. 329–331 for additional information on the initial success of Operation Challenge 4.

13 See Anthony H. Cordesman and Abraham R. Wagner, *The Lessons of Modern War (Volume 1): The Arab–Israeli Conflicts, 1973–1989* (Boulder, CO: Westview Press, 1989), p. 83. For a compact and insightful treatment of the aerial dimension of the Yom Kippur War, see pp. 73–102.

14 Cordesman and Wagner, *The Lessons of Modern War*, p. 83. Some accounts, it should be pointed out, claim that Operation Model 5 resulted in the destruction of one SAM battery. See Shmuel L. Gordon, "Air Superiority in the Israel–Arab Wars, 1967–1982," p. 146 in John Andreas Olsen (ed.), *A History of Air Warfare* (Washington, D.C.: Potomac Books, Inc., 2010). In either case, the operation ended in abject failure.

15 Cordesman and Wagner, *The Lessons of Modern War*, p. 90. Also see the October 22, 1973 memorandum of conversation between Kissinger and Israeli military and political leaders in Political Affairs and Relations, Arab–Israeli Conflict (27–14), Record Group 59, Subject–Numeric Files 70–73, United States National Archives, College Park, Maryland. In this document, IAF commander Major General Benjamin Peled is quoted to the effect that Israel had lost more than 100 aircraft in the war to this point in time (i.e., a few days before the final cease-fire agreement took effect). Parenthetically, on a per sortie basis, the overall IAF loss rate in the 1973 war was lower than the overall loss rate in the 1967 war. Though the IAF lost approximately 2.3 times as many aircraft in the Yom Kippur War, it flew about 3.5 times as many sorties in this conflict.

16 The official IAF tally claims 277 Arab aircraft shot down in air-to-air combat during the Yom Kippur War at a cost of only five Israeli aircraft. Many unofficial sources adopt somewhat higher figures for the

number of Arab aircraft shot down, as well as the number of Israeli aircraft lost in air battles.

17 Cordesman and Wagner, *The Lessons of Modern War*, p. 96.

18 Ibid., p. 90.

19 van Creveld, "Israel," p. 181.

20 Regardless of its name, the Bar-Lev Line was not intended to hold off a large-scale Egyptian thrust across the Suez Canal. It consisted of a series of quite small and widely dispersed strongholds meant simply to protect Israeli troops from Egyptian army shelling and to provide the IDF with observation posts on the front line.

21 The IAF suffered about half of its total aircraft losses during the first phase of the war. See van Creveld, "Israel," p. 182.

22 On the growth in the IAF's firepower between the wars, see ibid., pp. 170–171, 184.

23 For the generally poor performance of Egypt's air-to-ground missiles, see Cordesman and Wagner, *The Lessons of Modern War*, p. 98.

24 The fear of Israeli airpower may also have provided Jordan with the diplomatic cover that it required to refrain from attacking Israel from its own territory. Instead, the Jordanian army—and, for that matter, the Iraqi army—deployed token units to the Golan front to fight alongside the Syrian army.

25 The following paragraphs are intended only to illustrate the limits of the IAF's contribution to the ground campaigns in 1967 and 1973. A detailed assessment of the relative effectiveness of airpower versus ground power in these wars falls outside the scope of this monograph.

26 Sinai front divisional commander Major General Ariel Sharon, for instance, opted to fight the crucial battle at Abu Agueila at night. Though the IAF did transport paratroopers behind Egyptian lines during this battle, the air force otherwise did not participate in it. On the Abu Agueila battle, see van Creveld, "Israel," p. 166.

27 Even highly knowledgeable observers have subscribed to this line of thinking. For a representative example, see the January 14, 1974 letter from General William E. DePuy, onetime head of the United States Army's TRADOC (Training and Doctrine) Command, to United States Army Chief of Staff General Creighton Abrams in the Orwin C. Talbott Papers, Deputy CG TRADOC, Arab–Israeli War (1973) Box, Letter from General DePuy to General Abrams Folder, United States Army Military History Institute, Carlisle Barracks, Pennsylvania.

28 The IAF lost about 20–25 aircraft in CAS and interdiction sorties during the Six-Day War. It lost about 70–75 aircraft in the same types of sortie during the Yom Kippur War. In light of the fact that the IAF flew roughly three times as many CAS and interdiction sorties in the

1973 conflict, not to mention that it flew them in the face of much more robust IADSs, its loss rate in this war for these types of mission would not seem to compare unfavorably with the one obtained in the 1967 conflict.

29 Some knowledgeable observers have given an affirmative answer to this question, while others have given a negative answer. For the former, see Steven J. Rosen and Martin Indyk, "The Temptation to Pre-empt in a Fifth Arab–Israeli War," *Orbis*, Vol. 20, No. 3 (Summer 1976), pp. 271–272. For the latter, see John R. Carter, *Airpower and the Cult of the Offensive* (Maxwell Air Force Base, AL: Air University Press, 1998), pp. 52–65.

30 For problems with the IAF's reconnaissance activities in the Yom Kippur War, see Cordesman and Wagner, *The Lessons of Modern War*, pp. 99–100.

2 Airpower and Attrition Warfare: The Israel Air Force in the 1969–70, 2006, and 2008–9 Wars

1 The most comprehensive public data base of Israeli victories and losses in air-to-air combat during the War of Attrition can be found in Shlomo Aloni's two-volume set: Shlomo Aloni, *Israeli Mirage and Nesher Aces* (Oxford: Osprey Publishing Ltd., 2004), pp. 82–84 and Shlomo Aloni, *Israeli F-4 Phantom II Aces* (Oxford: Osprey Publishing Ltd., 2004), p. 86. For Israeli air victories in this period, see Aloni, *Israeli Mirage and Nesher Aces*, pp. 82–83.

2 Yaacov Bar-Siman-Tov, *The Israeli–Egyptian War of Attrition, 1969–1970: A Case Study of Limited Local War* (New York: Columbia University Press, 1980), p. 45. For a thorough review of Israeli special forces operations during the War of Attrition, see W. Andrew Terrill, "The Nature and Value of Commando Operations During the Egyptian–Israeli War of Attrition," *Small Wars & Insurgencies*, Vol. 8, No. 2 (Autumn 1997), pp. 16–34.

3 Bar-Siman-Tov, *The Israeli–Egyptian War of Attrition*, pp. 69–70.

4 Ibid., p. 87.

5 Ibid., pp. 90–102.

6 Aloni, *Israeli Mirage and Nesher Aces*, p. 83; Aloni, *Israeli F-4 Phantom II Aces*, p. 86; and Bar-Siman-Tov, *The Israeli–Egyptian War of Attrition*, pp. 90–102.

7 For accounts of Israel's deep-penetration bombing campaign, see Bar-Siman-Tov, *The Israeli–Egyptian War of Attrition*, pp. 117–173; Yaacov Bar-Siman-Tov, "The Myth of Strategic Bombing: Israeli Deep-Penetration Air Raids in the War of Attrition, 1969–1970," *Journal of Contemporary History*, Vol. 19, No. 3 (1984), pp. 549–570; and Avi Shlaim and Raymond Tanter, "Decision, Process, Choice,

and Consequences: Israel's Deep-Penetration Bombing in Egypt, 1970," *World Politics*, Vol. 30, No. 4 (July 1978), pp. 483–516.

8 For the number of sorties flown and range of targets struck by the IAF, see the summary of the War of Attrition on the air force's web site, *www.iaf.org.il.*

9 For the Soviet deployment to Egypt, see Efraim Karsh, *The Cautious Bear: Soviet Military Engagement in Middle Eastern Wars in the post-1967 Era* (Boulder, CO: Westview Press, 1987), pp. 16, 64–65, 77.

10 One IAF air strike, to cite a single example, resulted in 13 Soviet deaths. On this strike, see the remarks of Major General (Ret.) Avihu Bin-Nun, a participant in the attack, in Simcha Salach (ed.), *Israel Air Force Milestones: Heroes and Legends* (Herzliya: The Israel Air Force Center, 2007), p. 16. On Egyptian and Soviet losses in air battles during this stage of the war, see Aloni, *Israeli Mirage and Nesher Aces*, pp. 83–84 and Aloni, *Israeli F-4 Phantom II Aces*, p. 86. On IAF aircraft losses, see Eliot A. Cohen and John Gooch, *Military Misfortunes: The Anatomy of Failure in War* (New York: The Free Press, 1990), p. 99.

11 For the number of sorties in the IAF air campaign, see David Makovsky and Jeffrey White, *Lessons and Implications of the Israel–Hizballah War: A Preliminary Assessment* (Washington, D.C.: The Washington Institute for Near East Policy, 2006), p. 50. For a considerably higher figure, see Itai Brun, "The Second Lebanon War, 2006," p. 298 in John Andreas Olsen (ed.), *A History of Air Warfare* (Washington, D.C.: Potomac Books, Inc., 2010). In addition to these sources, others that contain much information on the IAF air campaign include: Anthony H. Cordesman, *Preliminary "Lessons" of the Israeli–Hezbollah War* (Washington, D.C.: Center for Strategic and International Studies, 2006); Benjamin S. Lambeth, *Air Operations in Israel's War against Hezbollah: Learning from Lebanon and Getting It Right in Gaza* (Santa Monica, CA: The Rand Corporation, 2011); and Noam Ophir, "Look Not to the Skies: The IAF vs. Surface-to-Surface Rocket Launchers," *Strategic Assessment*, Vol. 9, No. 3 (November 2006).

12 Makovsky and White, *Lessons and Implications*, p. 50.

13 Collectively speaking, the IAF's counterbattery strikes against Hizbullah rocket launchers are considered strategic attacks in the context of this monograph, because the objective of these strikes was clearly to defend the Israeli home front, surely a wartime goal of immense importance to Israel.

14 Makovsky and White, *Lessons and Implications*, p. 49.

15 On the destruction of this C^2 facility, see Cordesman, *Preliminary "Lessons,"* pp. 4, 16.

16 Ophir, "Look Not to the Skies."

17 Brun, "The Second Lebanon War," p. 317; Cordesman, *Preliminary "Lessons,"* p. 4; and Makovsky and White, *Lessons and Implications*, p. 49.
18 Ophir, "Look Not to the Skies."
19 Cordesman, *Preliminary "Lessons,"* p. 22; Makovsky and White, *Lessons and Implications*, p. 49; and Ophir, "Look Not to the Skies."
20 For a comprehensive (if controversial at points) treatment of Hizbullah rocket attacks on Israel, see Uzi Rubin, *The Rocket Campaign Against Israel During the 2006 Lebanon War* (Ramat Gan: The Begin–Sadat Center for Strategic Studies, 2007). For the number of rockets landing in Israel over the span of the war, see p. 10.
21 On Hizbullah's use of human shields in an attempt to protect itself against Israeli counterstrikes, see Reuven Erlich, *Hezbollah's Use of Lebanese Civilians as Human Shields* (Gelilot: Intelligence and Terrorism Information Center at the Center for Special Studies, 2006).
22 Cordesman, *Preliminary "Lessons,"* pp. 4–5.
23 Ibid., p. 22.
24 Makovsky and White, *Lessons and Implications*, p. 49.
25 During a June 25, 2007 visit to the Israel Air Force Center (IAFC), Herzliya, Israel, the author heard a young air force special forces "operator" describe his role in a number of harrowing casualty evacuation sorties.
26 For general accounts of Operation Cast Lead, including the IAF's part in hostilities, see Amir Kulick, "'Lebanon Lite': Lessons from the Operation in Gaza and the Next Round Against Hizbollah," *Military and Strategic Affairs*, Vol. 1, No. 1 (April 2009), pp. 51–66; Lambeth, *Air Operations in Israel's War against Hezbollah*; Matt M. Matthews, "Hard Lessons Learned," pp. 5–44 in Scott C. Farquhar (ed.), *Back to Basics: A Study of the Second Lebanon War and Operation Cast Lead* (Ft. Leavenworth, KS: United States Combat Studies Institute, 2009); and Abe F. Merraro, "The Tactics of Operation Cast Lead," pp. 83–102 in Farquhar, *Back to Basics*.
27 Again, the IAF's counterbattery strikes are considered strategic attacks for the purpose of this monograph, as their aim was to defend the Israeli home front.
28 Kulick, "'Lebanon Lite,'" p. 55.
29 Ibid., p. 55.
30 Matthews, "Hard Lessons Learned," p. 30 and Amir Rapoport, *The IDF and the Lessons of the Second Lebanon War* (Ramat Gan: The Begin–Sadat Center for Strategic Studies, 2010), p. 23.
31 Matthews, "Hard Lessons Learned," p. 30.
32 Cordesman, *Preliminary "Lessons,"* pp. 18–19.

33 On Hizbullah's drone sorties, see Rubin, *The Rocket Campaign Against Israel*, pp. 21–22.
34 Bar-Siman-Tov, *The Israeli–Egyptian War of Attrition*, p. 121.
35 Barbara Opall-Rome, "Inteview with Maj. Gen. Ido Nehushtan, Commander, Israel Air and Space Force," *DefenseNews*, August 15, 2011. Accessed at *www.defensenews.com*.
36 Hamas (and its allies) reportedly suffered 700–800 dead in the fighting. For the organization's own admission about its losses, see the November 3, 2010 bulletin issued by the Intelligence and Terrorism Information Center in Israel, which can be found at *www.terrorism-info.org.il*.

3 Airpower, Counterinsurgency, Special Operations, and Humanitarian Operations: The Israel Air Force between Arab–Israeli Wars

1 On the air force's participation in the Karameh raid, see Eliezer Cohen, *Israel's Best Defense: The First Full Story of the Israeli Air Force* (New York: Orion Books, 1993), pp. 259–262.
2 For the scale of PLO losses during the late 1960s, see Gunther E. Rothenberg, "Israeli Defence Forces and Low-Intensity Operations," p.65 in David Charters and Maurice Tugwell (eds.), *Armies in Low-Intensity Conflict: A Comparative Analysis* (London: Brassey's Defence Publishers, 1989).
3 The IAF's part in this round of counterinsurgency warfare is described in Cohen, *Israel's Best Defense*, pp. 436–444 and Rothenberg, "Israeli Defence Forces," pp. 68–70.
4 The IAF's campaign against Hizbullah during these years is reviewed in Shmuel L. Gordon, *The Vulture and the Snake: Counter-Guerrilla Air Warfare, the War in Southern Lebanon* (Ramat Gan: The Begin–Sadat Center for Strategic Studies, 1998).
5 For this incident, see Benjamin Armstrong, "Precision Approaches: Leadership Targeting and the Helicopter as a Strategic Strike Asset in Small Wars," *Defense & Security Analysis*, Vol. 25, No. 3 (September 2009), p. 277.
6 Anecdotal evidence on the effectiveness of the IAF's targeted attacks can be found in Armstrong, "Precision Approaches," pp. 277–278. For the number of terrorists eliminated by targeted attacks up to the autumn of 2005, see Alan M. Dershowitz, *Preemption: A Knife That Cuts Both Ways* (New York: W. W. Norton & Company, 2006), p. 315 (note 58). Approximately one-third of these terrorists were killed in aerial strikes. Also see Justus Reid Weiner, *Targeted Killings and Double Standards* (Jerusalem: Jerusalem Center for Public Affairs, 2012), which contains an exhaustive list of Israeli targeted attacks

through the end of 2008. This study and its statistical appendices can be found at the Jerusalem Center for Public Affairs web site, *www.jcpa.org*.

7 For an indication of the effectiveness of helicopter gunships in the interdiction and CAS roles during the second intifada, see Sergio Catignani, *Israel, Counter-insurgency, and the* Intifadas*: Dilemmas of a Conventional Army* (London: Routledge, 2008), p. 131 and Ra'anan Weiss, *AH-64 A/D Peten and Saraf in IAF Service* (Bat Hefer: Isradecal, 2011), p. 14.

8 During a briefing at Nevatim Air Force Base, Israel, on June 26, 2008, the base commander, Colonel A, offered implicit confirmation that the air force has also been involved in still-classified special operations. In remarks about the IAF's transport fleet, which was then in the process of relocating to this base, he mentioned that its aircraft had been to "some places that we cannot talk about."

9 For a recent account of the IAF raid, see Amos Perlmutter, Michael I. Handel, and Uri Bar-Joseph, *Two Minutes over Baghdad*, Second Expanded Edition (London: Routledge, 2003). For the perspective of one of the attacking pilots, see Iftach Spector, *Loud and Clear: The Memoir of an Israeli Fighter Pilot* (Minneapolis, MN: Zenith Press, 2009), pp. 385–416.

10 Israel has never officially taken credit for the raid, so details about how the IAF accomplished it are purely speculative at this point in time. A plausible reconstruction of the attack appears in Richard A. Clarke and Robert K. Knake, *Cyber War: The Next Threat to National Security and What To Do About It* (New York: Ecco, 2010), pp. 1–8.

11 Ibid., pp. 6–7.

12 The strike on PLO targets in Tunis is chronicled in Cohen, *Israel's Best Defense*, pp. 482–488.

13 For a recent account of the Entebbe raid, including the air force's part in it, see Simon Dunstan, *Israel's Lightning Strike: The Raid on Entebbe, 1976* (Oxford: Osprey Publishing Ltd., 2009).

14 Briefing on the history and role of Unit 669 by Captain Y, one of its "operators," at Tel Nof Air Base, Israel, on June 24, 2007. Y also screened some footage of Unit 669 rescue operations carried out in the 2006 Second Lebanon War.

15 During a talk at the Israel Air Force Center (IAFC), Herzliya, Israel, on June 24, 2008, Captain O, a UH-60 Blackhawk pilot, described a particularly harrowing 669 operation during the Second Lebanon War to evacuate wounded soldiers still under fire, and for which he received a decoration. Parenthetically, 669 does not have its own helicopters; they are assigned to the unit whenever necessary, and O was one of the pilots who worked often with the unit.

16 Basic information on Shaldag can be found on the IAF and IDF web sites, *www.iaf.org.il* and *www.idf.il*, respectively.

17 On the purpose of the IDF depth corps, see Ronen Cohen, "Tactical vs. Strategic Raid," *IsraelDefense*, Vol. 1, No. 6 (January–February 2012), pp. 38–41.

18 At a presentation in Tel Aviv, Israel, on June 23, 2007, Brigadier General (Res.) Asaf Agmon, who participated as a Hercules pilot in these clandestine flights into Sudan, described the military aspects of the operation in considerable detail and screened some still-classified footage of refugees being shepherded into a C-130.

19 It should be noted that Operation Moses did not consist solely of an airlift from Sudan. An indeterminate number of Ethiopian Jews arrived in Israel via a sealift and other means, but most got there by aircraft.

20 Brigadier General Agmon served as the overall IAF commander of the operation. Lieutenant General Benny Gantz, a future IDF Chief of Staff, led Shaldag at the time of the airlift; therefore, he served as commander of the land forces component of the operation. See Stephen Spector, *Operation Solomon: The Daring Rescue of the Ethiopian Jews* (New York: Oxford University Press, 2006) for a general account of the airlift.

21 Briefing on the history and role of Unit 669 by Captain Y at Tel Nof Air Base, Israel, on June 24, 2007.

22 The information in this section of the chapter can be found on the Israel Ministry of Foreign Affairs web site, *www.mfa.gov.il*, and the IDF web site, *www.idf.il*.

23 See Helen Davis and Douglas Davis, *Israel in the World: Changing Lives Through Innovation* (London: Weidenfeld & Nicolson, 2005), p. 188.

4 An Airborne Revolution in Military Affairs: Unmanned Aerial Vehicles in Israel Air Force Service

1 A concise description of Israeli drone operations in the Yom Kippur War is found in John F. Kreis, "Unmanned Aircraft in Israeli Air Operations," *Air Power History*, Vol. 37, No. 4 (Winter 1990), p. 46.

2 For concise descriptions of Israeli drone operations before and during the Lebanon War, see Kreis, "Unmanned Aircraft," pp. 47–49; Benjamin S. Lambeth, *Moscow's Lessons from the 1982 Lebanon Air War* (Santa Monica, CA: The Rand Corporation, 1984), pp. 4–8; and Ralph Sanders, "Israeli Military Innovation: UAVs," *Joint Forces Quarterly*, No. 33 (Winter 2002–3), p. 115. Until rather recently, it is appropriate to mention here, the IDF's military intelligence arm, A'MAN, also carried out drone sorties. The relationship of this UAV force to the IAF's own, as well as its specific role(s), has not been disclosed publicly.

3 The Tadmit is sometimes referred to as a bomb; however, because it had both a guidance system and a rocket motor, this monograph considers it to be a missile.
4 For this claim, see Richard A. Gabriel, *Operation Peace for Galilee: The Israeli–PLO War in Lebanon* (New York: Hill and Wang, 1984), p. 99. Israel has since produced a highly sophisticated, longer-range, "hunter–killer" follow-on to the Harpy called Harop. The IAF also has in its arsenal an air-launched "loitering" weapon system—that is, a weapon system that can linger over a target area for a period of time—known as Delilah.
5 Briefing by UAV operator Captain G of 200 "First UAV" Squadron, Palmachim Air Force Base, Israel, on June 22, 2009.
6 See Sanders, "Israeli Military Innovation," p. 117. Boost phase refers to the moments just after lift off, as the missile is accelerating away from the launch vehicle.
7 Israel has refused either to confirm or deny persistent media reports that the IAF uses armed drones for attack missions. When asked whether the IAF possessed such drones, UAV squadron commander Lieutenant Colonel N of 200 "First UAV" Squadron pointedly refrained from answering the question at a briefing at Palmachim Air Force Base, Israel, on June 23, 2008.
8 Ibid. Foreign officers, according to Lieutenant Colonel N, visit Palmachim on a regular basis to learn from the IAF.
9 For the role of Israeli UAVs in southern Lebanon in this period, see Shmuel L. Gordon, *The Vulture and the Snake: Counter-guerrilla Air Warfare, the War in Southern Lebanon* (Ramat Gan: The Begin–Sadat Center for Strategic Studies, 1998), p. 57.
10 To get a sense of drone participation in these conflicts, see, for example, Arie Egozi, "Israel Praises UAV Abilities During Operation Change of Direction Anti-Hezbollah Campaign," *Flight International*, August 29, 2006. Accessed at *www.flightglobal.com* in the UAV section of the site.
11 Concise descriptions of Israel's counter-rocket air campaign are found in Itai Brun, "The Second Lebanon War, 2006," pp. 297–323 in John Andreas Olsen (ed.), *A History of Air Warfare* (Washington, D.C.: Potomac Books, Inc., 2010); Anthony H. Cordesman, *Preliminary "Lessons" of the Israeli–Hezbollah War* (Washington, D.C.: Center for Strategic and International Studies, 2006); Benjamin S. Lambeth, *Air Operations in Israel's War against Hezbollah: Learning from Lebanon and Getting It Right in Gaza* (Santa Monica, CA: The Rand Corporation, 2011); David Makovsky and Jeffrey White, *Lessons and Implications of the Israel–Hizballah War: A Preliminary Assessment* (Washington, D.C.: The Washington Institute for Near East Policy, 2006); and Noam

Ophir, "Look Not to the Skies: The IAF vs. Surface-to-Surface Rocket Launchers," *Strategic Assessment*, Vol. 9, No. 3 (November 2006).

12 The author viewed a nonclassified video clip of one of these roof-knocker missiles in action during a 2012 visit to the IAF.

13 On drones as a cheap force multiplier in IAF eyes, see Arie Egozi, "Israel Broadens UAV Use with Advanced Designs," *Flight International*, February 11, 2008. Accessed at *www.flightglobal.com* in the UAV section of the site.

14 See, for example, the IAF order of battle at *www.scramble.nl*. The air force does not release official figures about the number of squadrons (manned or unmanned), or about the number of machines per squadron, in its order of battle; therefore, the following figures are estimates.

15 See Arie Egozi, "Israel Broadens UAV Use."

16 To get a sense of the Fisher Institute's research mandate, see its web site at *www.fisherinstitute.co.il/eng*. Perhaps the institute's most high-profile UAV-related project to date is the conceptualization of a medical evacuation drone that would transport wounded soldiers from "hot" battlefields where manned helicopters would be in harm's way.

17 See Barbara Opall-Rome, "Israel AF Hones Manned–UAV Mix," *DefenseNews*, July 7, 2008. Accessed at *www.defensenews.com*. For an insightful discussion of the technical and operational complexities of drone employment in IAF operations see Asaf Agmon and Tal Inbar, "UAVs Heading Where?: Future Trends in the Development of Unmanned Aerial Vehicles and Their Operational Use," *Strategic Analysis* (February 2006).

18 Indeed, the IAF already considers drones to be "substitutes" for satellites under certain circumstances. See Egozi, "Israel Broadens UAV Use."

19 A country with a second strike capability is one whose arsenal of nuclear (or biological or chemical) weapons can survive an opponent's first strike with nuclear (or biological or chemical) weapons, permitting it to respond with a devastating counterstrike of its own.

20 For Major General (Ret.) Ido Nehushtan's remarks about drones, see Amir Rapoport, "Missions on the Edge of the Horizon," *IsraelDefense*, Vol. 1, No. 3 (June–July 2011), p. 15.

5 Ground-based Air Defense and Space-based Reconnaissance: Other War-related Responsibilities of the Israel Air Force

1 A brief review of Israel's anti-aircraft defenses, based on official IDF sources, can be found on the Internet at *www.jewishvirtuallibrary.org* in the section of the site devoted to the IAF.

2 See the tables in Anthony H. Cordesman and Abraham R. Wagner,

The Lessons of Modern War (Volume 1): The Arab–Israeli Conflicts, 1973–1989 (Boulder, CO: Westview Press, 1989), pp. 75–76.

3 See *www.jewishvirtuallibrary.org.*

4 Ibid.

5 Ibid. See also the table in Cordesman and Wagner, *The Lessons of Modern War*, p. 91.

6 See the tables in Cordesman and Wagner, *The Lessons of Modern War*, pp. 75–76 and 78.

7 The lower figure is cited by official IDF sources, while the higher figure is cited by two respected military analysts. See *jewishvirtuallibrary.org* and Cordesman and Wagner, *The Lessons of Modern War*, p. 89, respectively.

8 See *www.jewishvirtuallibrary.org* and Cordesman and Wagner, *The Lessons of Modern War*, pp. 185 and 193.

9 For a review of the Patriot SAM's abysmal performance in the Gulf War, see Theodore Postol, "Lessons of the Gulf War Experience with the Patriot," *International Security*, Vol. 16, No. 3 (1991/2), pp. 119–171.

10 See Uzi Rubin, *The Missile Threat from Gaza: From Nuisance to Strategic Threat* (Ramat Gan: The Begin–Sadat Center for Strategic Studies, 2011), p. 41. The author had the opportunity to tour this battery during a visit to Palmachim Air Force Base, Israel, on June 23, 2008. Captain G, the battery's operations officer, delivered a briefing on the system's capabilities.

11 See Nir Dvori, "An Excess of Technological Risks," *IsraelDefense*, Vol. 1, No. 3 (June–July 2011), p. 56.

12 See Yael Livnat, "The Future of Air Defense: Magic Wand and Arrow III Systems," Department of Israel Defense Forces Spokesperson, July 8, 2012. Accessed at the official IDF web site, *www.idf.il.* The upgraded interceptor, in other words, will be able to destroy incoming ballistic missiles at longer ranges with even greater accuracy than the current model.

13 Conversation in Jerusalem, Israel, on June 24, 2009 with a female sergeant attached to a Patriot SAM battery stationed near Dimona.

14 Livnat, "The Future of Air Defense."

15 On the IDF's foot-dragging, see Rubin, *The Missile Threat from Gaza*, pp. 43–48.

16 Ibid.

17 For the various estimates, see Rubin, *The Missile Threat from Gaza*, p. 69 and Yaacov Katz, "Iron Dome Successful in Downing 75% of Rockets," *Jerusalem Post*, December 31, 2011. Accessed at *www.jpost.com.*

18 See "New Interceptor Missile To Be Used by Iron Dome Defense

System," Department of Israel Defense Forces Spokesperson, July 25, 2012. Accessed at the IDF web site, *www.idf.il.*

19 On the American refusal, see Deganit Paikowsky, "Israel in Space," *Strategic Analysis* (August 2006), pp. 11–29.

20 For overviews of Israel's satellite program, see Efrat Cohen and Moriya Ben Josef, "Not Only Satellites: Israel's Place in Space," *IsraelDefense*, Vol. 1, No. 3 (June–July 2011), pp. 36–38; Amir Rapaport, "Israel's Space Program," *IsraelDefense*, Vol. 1, No. 6 (January–February 2012), pp. 34–37; and Danny Shalom, "Israel's Space Program Is at a Crossroads," *IsraelDefense*, Vol. 1, No. 5 (October–November 2011), pp. 20–22. Also see Adam Baddeley, "Israel Exploits Space Technologies, Capabilities," *Signal Magazine* (March 2011). This article can be found at *www.afcea.org.*

21 Israel's fleet of Jericho ballistic missiles, which is reportedly armed with nuclear warheads, is based on the Shavit booster rocket. It is unclear whether the IAF has control over Israel's ballistic missile fleet—or, for that matter, over the state's arsenal of air-dropped bombs—which is why the issue of nuclear weapons is not addressed in this book.

22 See Shalom, "Israel's Space Program," p. 22. Also see Yoram Ilan-Lipovski, *Microsatellites and Airborne Launching: New Goals in Space* (Herzliya: Fisher Institute of Air and Space Strategic Studies, ND). Satellite expert Dr. Tal Inbar commented on this capability as well at a June 25, 2012 briefing at the Israel Air Force Center (IAFC), Herzliya, Israel.

23 Rapaport, "Israel's Space Program," p. 36.

6 The Infrastructure of the Israel Air Force: Bases and Squadrons

1 The author paid a short visit to Israel Air Force (IAF) headquarters on June 29, 2011 as part of an Israel Air Force Center (IAFC) delegation. For security reasons, however, the group was restricted to a couple of public areas on the lower floors of the building, and did not have an opportunity to tour the underground command center. For the existence of this center, see Danny Shalom and Amir Rapoport, "50 Seconds that Changed the World" *IsraelDefense*, Vol. 1, No. 3 (June–July 2011), pp. 72–79, especially the photo and caption on p. 73.

2 For a fascinating account of bird–aircraft coexistence in Israeli skies, see Yossi Leshem and Ofer Bahat, *Flying with the Birds* (Tel Aviv: Chemed Books, 1999).

3 See "A Rubber Punch" in the online edition of the *Israel Air Force Magazine*, which can be found on the IAF web site, *www.iaf.org.il.*

4 For recent examples of IAF training in Italian and Greek airspace, respectively, see Lya Shanel, "Bonjourno Uvda" and Gal Goldstein, "Top of the Olympus" in the online edition of the *Israel Air Force*

Magazine. The air force has also taken part in the Red Flag exercises hosted by the United States Air Force (USAF) in the Nevada desert.

5 At least one fighter-bomber squadron, however, flew from Israel's main civilian airport during the Six-Day War.

6 All of the IAF's fighter-bomber squadrons are capable of fulfilling each of the air force's combat roles—air superiority, close air support (CAS), interdiction, and strategic attack; nevertheless, despite the fact that the air force has always favored the acquisition of multirole aircraft, each squadron also has a primary task.

7 Major Y, deputy commander of the squadron, provided this figure during a briefing at Ramat David Air Force Base, Israel, on July 1, 2012.

8 The author has been personally able to view the squadron's "kill board" on several visits to Tel Nof Air Force Base. Intriguingly, on a June 25, 2012 visit to the unit, one of the young pilots pointed out that the squadron's aircraft possess the capability to lock each other's missiles on to targets, a kind of intranet in the sky. Whether the IAF's other F-15 squadrons, not to mention its F-16 units, possess the same capability is unknown.

9 During a June 25, 2012 briefing at 118 squadron, Captain I politely declined to get into specifics about the unit's operations.

10 The author had the opportunity to tour both Depot 22 and MANAT as part of an IAFC delegation on June 27, 2010. The delegation received briefings from Lieutenant Colonel I, the chief maintenance officer, and Lieutenant Colonel A, the chief test pilot, on the functions of Depot 22 and MANAT, respectively.

11 Conversation with Major I, deputy commander of the squadron, at the IAFC, Herzliya, Israel, on June 28, 2010.

12 The author heard a rather poignant "father-and-son" story from Captain O, a young C-130 pilot with 103 squadron, during a conversation at the IAFC on June 25, 2012. O's father, a member of Sayeret Matkal, flew to Entebbe as part of the Israeli rescue force aboard one of the aircraft that his son would later pilot.

13 The IAF also houses some out-of-service B707s and C-130s at Ben-Gurion Airport.

14 The author visited Sde Dov Air Force Base with an IAFC delegation on June 28, 2011. The briefing provided by several pilots of one of these squadrons, not surprisingly, did not address the intelligence side of their duties.

15 See Amir Rapoport, "Missions on the Edge of the Horizon," *IsraelDefense*, Vol. 1, No. 3 (June–July 2011), pp. 16–17.

16 Conversation with Lieutenant O, passive defense officer, at Ramat David Air Force Base, Israel, on June 27, 2007.

17 The author has met with some of these young men and women during repeated visits to the IAFC.

Conclusion: The Past and Future Contributions of Airpower to Israeli National Security

1 This point was made to the author in a conversation at the Israel Air Force Center (IAFC), Herzliya, Israel, on June 28, 2010 by Major I, an air force fighter-bomber pilot.
2 For two views of the IAF's decision to acquire this aircraft, see Gur Laish, "Israel and the F-35: A Look Beyond Costs and Politics," *Strategic Assessment*, Vol. 13, No. 4 (January 2011), pp. 7–19 and Yiftah Shapir, "The F-35 Deal: An Enlightened Purchase?," *Strategic Assessment*, Vol. 13, No. 4 (January 2011), pp. 21–38. Laish is an enthusiastic proponent of the F-35; Shapir is less enamored of the aircraft.
3 Brigadier General (Ret.) Ephraim Segoli, an expert on airpower in asymmetrical warfare, remarked upon the IAF's traditional reluctance to undertake the CAS role in a talk at an Israel Air Force Center Foundation (IAFCF) event on Long Island, New York, on November 7, 2010. For the integration of IAF officers into land forces units, see, for example, Arie Egozi, "IDF's Air, Ground Forces Seek Increased Cooperation," *IsraelDefense*. Accessed at the *IsraelDefense* web site, *www.israeldefense.com*.
4 See Yoav Tsuk, "A Winning Combination," *www.iaf.org.il*.
5 Amir Rapoport, "Missions on the Edge of the Horizon," *IsraelDefense*, Vol. 1, No. 3 (June–July 2011), p. 12.

Appendix: The Historical Evolution of the Israel Air Force's Fighter-Bomber Inventory

1 For the tortuous negotiations leading to the initial sales of the A-4 Skyhawk (in 1966) and the F-4 Phantom (in 1968), see David Rodman, *Arms Transfers to Israel: The Strategic Logic Behind American Military Assistance* (Brighton and Portland, OR: Sussex Academic Press, 2007).
2 The Lavi relied heavily on American funding for its development. Once this funding was cut off, the project died swiftly in the late 1980s.

Select Bibliography

Articles

Agmon, Asaf and Tal Inbar, "UAVs Heading Where?: Future Trends in the Development of Unmanned Aerial Vehicles and Their Operational Use," *Strategic Analysis* (February 2006).

Armstrong, Benjamin, "Precision Approaches: Leadership Targeting and the Helicopter as a Strategic Strike Asset in Small Wars," *Defense & Security Analysis*, Vol. 25, No. 3 (September 2009): 271–284.

Bar-Siman-Tov, Yaacov, "The Myth of Strategic Bombing: Israeli Deep-Penetration Air Raids in the War of Attrition, 1969–1970," *Journal of Contemporary History*, Vol. 19, No. 3 (1984): 549–570.

Kreis, John F., "Unmanned Aircraft in Israeli Air Operations," *Air Power History*, Vol. 37, No. 4 (Winter 1990): 46–50.

Kulick, Amir "'Lebanon Lite': Lessons from the Operation in Gaza and the Next Round Against Hizbollah," *Military and Strategic Affairs*, Vol. 1, No. 1 (April 2009): 51–66.

Laish, Gur, "Israel and the F-35: A Look Beyond Costs and Politics," *Strategic Assessment*, Vol. 13, No. 4 (January 2011): 7–19.

Ophir, Noam, "Look Not to the Skies: The IAF vs. Surface-to-Surface Rocket Launchers," *Strategic Assessment*, Vol. 9, No. 3 (November 2006).

Pollack, Kenneth M., "Air Power in the Six-Day War," *The Journal of Strategic Studies*, Vol. 28, No. 3 (June 2005): 471–503.

Rodman, David, "Beyond the Battlefield: The Humanitarian Operations of the Israel Defense Forces," *Midstream*, Vol. 52, No. 1 (January–February 2006): 13–14.

Rodman, David, "Israel's National Security Doctrine: An Introductory Overview," *MERIA Journal*, Vol. 5, No. 3 (September 2001): 71–86.

Rodman, David, "Israel's National Security Doctrine: An Appraisal of the Past and a Vision of the Future," *Israel Affairs*, Vol. 9, No. 4 (Summer 2003): 115–140.

Rodman, David, "The Israel Air Force in the 1967 and 1973 Wars: Revisiting the Historical Record," *Israel Affairs*, Vol. 16, No. 2 (April 2010): 219–233.

Rosen, Steven J. and Martin Indyk, "The Temptation to Pre-empt in a

Fifth Arab–Israeli War," *Orbis*, Vol. 20, No. 3 (Summer 1976): 265–285.

Sanders, Ralph, "Israeli Military Innovation: UAVs," *Joint Forces Quarterly*, No. 33 (Winter 2002–3): 114–118.

Shapir, Yiftah, "The F-35 Deal: An Enlightened Purchase?," *Strategic Assessment*, Vol. 13, No. 4 (January 2011): 21–38.

Shlaim, Avi and Raymond Tanter, "Decision, Process, Choice, and Consequences: Israel's Deep-Penetration Bombing in Egypt, 1970," *World Politics*, Vol. 30, No. 4 (July 1978): 483–516.

Terrill, W. Andrew, "The Nature and Value of Commando Operations During the Egyptian–Israeli War of Attrition," *Small Wars & Insurgencies*, Vol. 8, No. 2 (Autumn 1997): 16–34.

Books and Studies

Aloni, Shlomo, *Israeli F-4 Phantom II Aces* (Oxford: Osprey Publishing Ltd., 2004).

Aloni, Shlomo, *Israeli Mirage and Nesher Aces* (Oxford: Osprey Publishing Ltd., 2004).

Aloni, Shlomo, *The June 1967 Six-Day War (Volume A): Operation Focus* (Bat Hefer: Isradecal, 2008).

Bar-Siman-Tov, Yaacov, *The Israeli–Egyptian War of Attrition, 1969–1970: A Case Study of Limited Local War* (New York: Columbia University Press, 1980).

Ben-Horin, Yoav and Barry Posen, *Israel's Strategic Doctrine* (Santa Monica, CA: The Rand Corporation, 1981).

Brecher, Michael, *Decisions in Crisis: Israel, 1967 and 1973* (Berkeley: University of California Press, 1980).

Carter, John R., *Airpower and the Cult of the Offensive* (Maxwell Air Force Base, AL: Air University Press, 1998).

Catignani, Sergio, *Israel, Counterinsurgency, and the* Intifadas*: Dilemmas of a Conventional Army* (London: Routledge, 2008).

Charters, David and Maurice Tugwell (eds.), *Armies in Low-Intensity Conflict: A Comparative Analysis* (London: Brassey's Defence Publishers, 1989).

Clarke, Richard A. and Robert K. Knake, *Cyber War: The Next Threat to National Security and What To Do About It* (New York: Ecco, 2010).

Cohen, Eliezer, *Israel's Best Defense: The First Full Story of the Israeli Air Force* (New York: Orion Books, 1993).

Cohen, Eliot A. and John Gooch, *Military Misfortunes: The Anatomy of Failure in War* (New York: The Free Press, 1990).

Cohen, Eliot A., Michael J. Eisenstadt, and Andrew J. Bacevich, *Knives, Tanks, and Missiles: Israel's Security Revolution* (Washington, D.C.: The Washington Institute for Near East Policy, 1998).

Cordesman, Anthony H., *Preliminary "Lessons" of the Israeli–Hezbollah War* (Washington, D.C.: Center for Strategic and International Studies, 2006).
Cordesman, Anthony H. and Abraham R. Wagner, *The Lessons of Modern War (Vol. 1): The Arab–Israeli Conflicts, 1973–1989* (Boulder, CO: Westview Press, 1989).
Davis, Helen and Douglas Davis, *Israel in the World: Changing Lives Through Innovation* (London: Weidenfeld & Nicolson, 2005).
Dershowitz, Alan M., *Preemption: A Knife That Cuts Both Ways* (New York: W. W. Norton & Company, 2006).
Dunstan, Simon, *Israel's Lightning Strike: The Raid on Entebbe, 1976* (Oxford: Osprey Publishing Ltd., 2009).
Dupuy, Trevor N., *Elusive Victory: The Arab–Israel Wars, 1947–1974* (New York: Random House, 1978).
Dupuy, Trevor N. and Paul Martell, *Flawed Victory: The Arab–Israeli Conflict and the 1982 War in Lebanon* (Fairfax, VA: Hero Books, 1986).
Erlich, Reuven, *Hezbollah's Use of Lebanese Civilians as Human Shields* (Gelilot: Intelligence and Terrorism Information Center at the Center for Special Studies, 2006).
Farquhar, Scott C. (ed.), *Back to Basics: A Study of the Second Lebanon War and Operation Cast Lead* (Ft. Leavenworth, KS: United States Combat Studies Institute, 2009).
Gabriel, Richard A., *Operation Peace for Galilee: The Israeli–PLO War in Lebanon* (New York: Hill and Wang, 1984).
Gordon, Shmuel L., *The Vulture and the Snake: Counter-Guerrilla Air Warfare, the War in Southern Lebanon* (Ramat Gan: The Begin–Sadat Center for Strategic Studies, 1998).
Handel, Michael, *Israel's Political–Military Doctrine* (Cambridge: Harvard University Press, 1973).
Herzog, Chaim, *The Arab–Israeli Wars: War and Peace in the Middle East from the War of Independence Through Lebanon* (New York: Random House, 1982).
Jackson, Robert, *The Israeli Air Force Story* (London: Tom Stacy, 1970).
Karsh, Efraim, *The Cautious Bear: Soviet Military Engagement in Middle Eastern Wars in the post-1967 Era* (Boulder, CO: Westview Press, 1987).
Lambeth, Benjamin, S., *Air Operations in Israel's War against Hezbollah: Learning from Lebanon and Getting It Right in Gaza* (Santa Monica, CA: The Rand Corporation, 2011).
Lambeth, Benjamin S., *Moscow's Lessons from the 1982 Lebanon War* (Santa Monica, CA: The Rand Corporation, 1984).
Levite, Ariel, *Offense and Defense in Israeli Military Doctrine* (Boulder, CO: Westview Press, 1989).

Luttwak, Edward N. and Dan Horowitz, *The Israeli Army* (New York: Harper and Row, 1975).

Makovsky, David and Jeffrey White, *Lessons and Implications of the Israel–Hizballah War: A Preliminary Assessment* (Washington, D.C.: The Washington Institute for Near East Policy, 2006).

Murray, Douglas J. and Paul R. Viotti (eds.), *The Defense Policies of Nations: A Comparative Study*, Third Edition (Baltimore: The Johns Hopkins University Press, 1994).

Murray, Williamson, MacGregor Knox, and Alvin Bernstein (eds.), *The Making of Strategy: Rulers, States, and War* (New York: Cambridge University Press, 1994).

Norton, Bill, *On the Edge: Aircraft of the Israel Air Force Since 1948* (Hinckley: Midland Publishing Ltd., 2002).

Olsen, John Andreas (ed.), *A History of Air Warfare* (Washington, D.C.: Potomac Books, Inc., 2010).

Perlmutter, Amos, Michael I. Handel, and Uri Bar-Joseph, *Two Minutes over Baghdad*, Second Expanded Edition (London: Routledge, 2003).

Rapoport, Amir, *The IDF and the Lessons of the Second Lebanon War* (Ramat Gan: The Begin–Sadat Center for Strategic Studies, 2010).

Rodman, David, *Arms Transfers to Israel: The Strategic Logic Behind American Military Assistance* (Brighton and Portland, OR: Sussex Academic Press, 2007).

Rubin, Uzi, *The Missile Threat from Gaza: From Nuisance to Strategic Threat* (Ramat Gan: The Begin–Sadat Center for Strategic Studies, 2011).

Rubin, Uzi, *The Rocket Campaign Against Israel During the 2006 Lebanon War* (Ramat Gan: The Begin–Sadat Center for Strategic Studies, 2007).

Salach, Simcha (ed.), *Israel Air Force Milestones: Heroes and Legends* (Herzliya: The Israel Air Force Center, 2007).

Spector, Iftach, *Loud and Clear: The Memoir of an Israeli Fighter Pilot* (Minneapolis, MN: Zenith Press, 2009).

Spector, Stephen, *Operation Solomon: The Daring Rescue of the Ethiopian Jews* (New York: Oxford University Press, 2006).

Tal, Israel, *National Security: The Israeli Experience* (Westport, CT: Praeger, 2000).

van Creveld, Martin, Steven L. Canby, and Kenneth S. Brower, *Air Power and Maneuver Warfare* (Maxwell Air Force Base, AL: Air University Press, 1994).

Weiner, Justus Reid, *Targeted Killings and Double Standards* (Jerusalem: Jerusalem Center for Public Affairs, 2012).

Weiss, Ra'anan, *AH-64 A/D Peten and Saraf in IAF Service* (Bat Hefer: Isradecal, 2011).

Yaniv, Avner, *Deterrence Without the Bomb: The Politics of Israeli Strategy* (Lexington, MA: Lexington Books, 1987).

Yonay, Ehud, *No Margin for Error: The Making of the Israeli Air Force* (New York: Pantheon Books, 1993).

Primary Sources

Articles on various American and Israeli web sites.
Briefings, e-mail communications, personal conversations, and speeches by retired and active Israel Air Force officers.
Various United States Government documents in the National Archives, College Park, Maryland and in the United States Army Military History Institute, Carlisle Barracks, Pennsylvania.

Index

About the Author

David Rodman is the author of two books about Israel: *Arms Transfers to Israel: The Strategic Logic Behind American Military Assistance* (Sussex Academic Press) and *Defense and Diplomacy in Israel's National Security Experience: Tactics, Partnerships, and Motives* (Sussex Academic Press). He has also published a number of articles in peer-reviewed academic journals, including *The Journal of Strategic Studies*, *Middle Eastern Studies*, *Israel Affairs*, *MERIA Journal*, *Chronicles Online Journal*, and *Israel Journal of Foreign Affairs*.